SIGNALS AND COMMUNICATION TECHNOLOGY

For other titles published in this series, go to
www.springer.com/series/4748

Luis M. Correia · Henrik Abramowicz ·
Martin Johnsson · Klaus Wünstel

Editors

Architecture and Design for the Future Internet

4WARD Project

Foreword by Joao Schwarz da Silva

 Springer

Editors
Prof. Luis M. Correia
Technical University of Lisbon
Instituto Telecomunicacoes
Instituto Superior Tecnico
Av. Rovisco Pais
1049-001 Lisbon,
Portugal
luis.correia@lx.it.pt

Dr. Henrik Abramowicz
Ericsson Research
Isafjordsgatan 14E
16480 Stockholm
Sweden

Dr. Martin Johnsson
Ericsson AB
Torshamnsgatan 23
16480 Stockholm
Sweden

Dr. Klaus Wünstel
Alcatel Lucent Bell Labs
Lorenzstrasse 10
70435 Stuttgart
Germany

ISSN 1860-4862
ISBN 978-90-481-9345-5   e-ISBN 978-90-481-9346-2
DOI 10.1007/978-90-481-9346-2
Springer Dordrecht Heidelberg London New York

Cover design: VTEX, Vilnius

Printed on acid-free paper

Springer is part of Springer Science+Business Media (www.springer.com)

Foreword

The development of computing and resource sharing as we have known them until recently is about to radically change course as its center of gravity is shifting with technologies, service architectures allowing for applications to migrate to the cloud. This shift from Web 2.0 to Web 3.0 will give rise to an Internet of services of unprecedented scope and scale. We are now entering a new phase of ICT driven innovation and growth based on the Internet of Services which more and more will be accessible through what could be called the Mobile and Wireless Web. Already today applications of wireless technology are a major driver of economic value in the EU economy. These are estimated at 250 bn€ or 2–3% of GDP and rising. In the coming five years it is expected that close to 7 billion users or the entire planet's population will have use of a mobile handset of which a great majority will be devices classified as smart-phones.

This is an unprecedented development exceeding the diffusion rates of technologies such as television or even pen and paper not only in terms of penetration and use but in its speed of take-up. We should expect an explosion of new applications with the potential to radically change the way in which we live and work. Examples are easy to cite: industrial and commercial applications in the supply chain, nomadic services for mobile workers, remote environmental monitoring or disaster and security systems that save lives by putting essential information into the hands of first responders, health and education services.

In such a remodelled world, new alliances will be created, new stakeholders will emerge, new modes of interaction will filter through into business practices, and new business models will proliferate. The Internet itself will no longer be a network of networks simply connecting computers and servers to become an Internet that connects "things" together: communicating devices by the billions, cars, machines of all sorts, household appliances, energy meters, windows, lights, etc. Around this new Internet will be borne a new economy of web based services and applications.

There are two key implications of this new Internet. First, this new world wide web of "things that think" will create a sensory network that will allow a leap forward in the human knowledge about the world we live in. It will lend itself to all sorts of new applications such as energy efficiency, health and welfare services, efficient transport and so on. If we do this well, there will be a massive improvement in our quality of life and sustainability, not just because of the services, not just because of the competitive advantage of being an earlier mover, but because European values of openness and democracy will define the form that the Internet takes.

Second we must liberate the economic potential of the single European market that is still locked up in fragmented national markets. In particular we must now strengthen the real economy by stimulating solid and sustainable business growth in high value goods and services that respond to real market needs. From the current period of uncertainty and as inevitable structural changes emerge it is essential to look for the growth opportunities in tomorrow's world. The industrial and research community gathered around the Future Internet Assembly, has certainly the talent and the capability to shape the future. All it takes is the ambition to overcome fragmented markets and the will to build on our strengths by creating open single markets for innovative goods and services and by going for innovation and change.

In creating the conditions that will allow Europe to benefit from the emerging economic opportunities, we must make sure that the Future Internet remains open. The key economic characteristic of the current Internet has been that it has created an unprecedentedly open platform for innovation and development of new services. We must keep this characteristic of openness by ensuring that open standards and eventually open-source software are the core of our actions. While the financial health of many companies worldwide is still based on proprietary models and gatekeeper business models, the world ahead of us will call for models whose economic basis offers a greater degree of liberty to the consumer or the enterprise.

As the Future Internet unfolds before us, the need will arise to move toward smarter and greener infrastructures. This is a big challenge, but also a great opportunity, because it will amplify the reach of the Internet to novel usages and industrial sectors. Indeed time has now come to go one step beyond what has been achieved so far. We must closely couple our Future Internet technology research and development with applications of high societal value such as health, urban mobility, energy grids or smart cities. In doing so, we will be able to provide an early "Internet response" to the many societal challenges with which we are confronted today.

Multiple regional initiatives are currently emerging in view of defining the future Internet. Japan and Korea have made public their ambitious u-Japan and u-Korea initiatives, China is supporting the domain through an ambitious and integrated industrial policy, in the US the GENI programme and facility is a key contributor to the debate on the future of the Internet. These initiatives are not all tackling the issue of the Internet evolution as part of their core objectives, but are certainly related to technological and socio-economic scenarios (ubiquity, connected devices) that will clearly need to be taken into account when addressing the Internet of Tomorrow.

From an EU perspective, it would be beneficial to build on these various initiatives and create the conditions that would bring about a closer complementarity and cooperation between all actors associated to the definition, testing and validation work. One of the main objectives of multilateral partnerships should be the emergence of global standards. Standards are indeed a key element to achieve interoperability and openness, two of the essential Internet characteristics that have contributed to its success. Indeed the ever growing multiplicity of players as well as the convergence of different sectors has lead to increased complexities in the standards making processes as illustrated by debates on IPR portfolios, as well as on the degree of openness, transparency and access.

Early co-operation and international partnerships on novel technologies are hence key to facilitate broader consensus, early agreements on standards by the key players while holding the promise to alleviate subsequent IPR disputes.

An important point to note is that the new economy created by the Internet is producing beyond a business revolution a unique opportunity to generate enormous environmental benefits, particularly if the right technological choices are made at the level of the infrastructure. In addition by reducing the amount of energy and materials consumed by business and by increasing overall productivity, the new Internet holds the promise to revolutionize the relation between economic growth and the environment.

It is in the above context that I have the pleasure to share with you my satisfaction as to the achievements of the EU R&D Project 4WARD. The book you are about to read, details the many unique contributions of the project to the development of a solid scientific basis for the Future Internet. Key amongst its many contributions are those relating to a new architecture framework where mobility, multi-homing and security become an intrinsic part of the network architecture rather than add-on solutions, hence allowing networks to bloom as a family of interoperable networks each complementing each other and each addressing individual requirements such as mobility, QoS, security, resilience, wireless transport and energy-awareness. 4WARD also addressed particularly well the question as to how virtualization can provide an opportunity to roll out new architectures, protocols, and services with network service providers sharing a common physical infrastructure. Tightly coupled to virtualization is network management, where 4WARD has broken new territory by advocating an approach where management functions come as embedded capabilities of devices. 4WARD has gone further than others by recognizing the paradigm shift brought about by the move from a node-centric age to an information-centric age.

The partners and scientific staff of 4WARD are to be congratulated for the work performed and for providing a perfect illustration of how Europe's commitment and creativity will enable the future.

Dr. Joao Schwarz da Silva
Former Director of DG-INFSO, European Commission

Preface

The current Internet is a tremendous commercial success and has become widely spread after having started as an academic research network to become a network for the everyday life for ordinary people. The Internet of today has its origins from the 70-ties, and was essentially simple but open for new applications and designed for the fixed network. It is however been increasingly challenged by the new transmission technologies based on radio and fiber, as well as by the new applications and media types that increasingly rely on overlays to make up for shortages in the core Internet architecture. In particularly, the even greater success of mobile networks has questioned the current Internet, which has reached a state of high complexity with regard to support of mobility, interoperability, configuration and management and vulnerability in an untrustworthy world.

The project 4WARD, started January 2008 and completed by June 2010, had the task to research on Architecture and Design for a Future Internet. The project took a clean slate research approach, which means that in its research it was not constrained by the current Internet. It does not mean however that the project favored a clean slate deployment, but rather saw a migration approach in how to apply its research results into the current Internet.

The project was partly EU funded under the EU Framework Programme 7 and consists of the 33 partners (see Appendix). There have been over 120 persons in the project, and for this reason it is not possible to list all that have contributed to the project and results. We would however like to acknowledge all for their valuable contributions. Further to that, we would like to acknowledge the help and support the project has experienced by the project officer Dr. Paulo de Sousa and the good collaboration we have had. The work of Daniel Sebastiao (IST, Lisbon) in the editing work is also acknowledged.

This book describes the salient results out of this project and covers not only technical results but deals also with socio-economic issues.

The Editors

Disclaimer

This book has been produced in the context of the 4WARD Project. The research leading to these results has received funding from the European Community's Seventh Framework Programme ([FP7/2007–2013]) under grant agreement n° 216041. All information in this document is provided "as is" and no guarantee or warranty is given that the information is fit for any particular purpose. The user thereof uses the information at its sole risk and liability. For the avoidance of all doubts, the European Commission has no liability in respect of this book, which is merely representing the authors view.

Contents

Contributors

Editors

Henrik Abramowicz Ericsson Research, Stockholm, Sweden

Pedro Aranda Gutiérrez Telefonica I+D, Madrid, Spain

Thorsten Biermann University of Paderborn, Paderborn, Germany

Anna Maria Biraghi Telecom Italia, Turin, Italy

Roland Bless Karlsruhe Institute of Technology, Karlsruhe, Germany

Jorge Carapinha PT Inovação, Aveiro, Portugal

Luis M. Correia IST/IT—Technical University of Lisbon, Lisbon, Portugal

Daniel Gillblad SICS—Swedish Institute of Computer Science, Stockholm, Sweden

Alberto Gonzalez Prieto KTH—Royal Institute of Technology, Stockholm, Sweden

Martin Johnsson Ericsson Research, Stockholm, Sweden

Holger Karl University of Paderborn, Paderborn, Germany

Denis Martin Karlsruhe Institute of Technology, Karlsruhe, Germany

Septimiu Nechifor Siemens, Brasov, Romania

Susana Perez Sanchez Tecnalia-Robotiker, Zamudio (Vizcaya), Spain

Jukka Salo Nokia Siemens Networks, Espoo, Finland

Göran Schultz Ericsson Research, Jorvas, Finland

Hagen Woesner Technical University of Berlin & EICT, Berlin, Germany

Klaus Wünstel Alcatel Lucent Bell Labs, Stuttgart, Germany

Martina Zitterbart Karlsruhe Institute of Technology, Karlsruhe, Germany

Other Contributors

Alexander Landau, Anders Eriksson, Andrei Bogdan Rus, Anghel Botos, Asanga Udugama, Avi Miron, Bengt Ahlgren, Björn Grönvall, Bogdan Tarnauca, Börje Ohlman, Carmelita Gorg, Chris Foley, Christian Dannewitz, Christian Tschudin, Christoph Werle, Daniel Horne, Daniel Sebastião, Djamal Zeghlache, Dominique Dudowski, Eric Renault, Fabian Wolff, Fabrice Guillemin, Fetahi Wuhib, Gabriel Lazar, Georgeta Boanea, Gerhard Hasslinger, Giorgio Nunzi, Gorka Hernando Garcia, Ian Marsh, Jim Roberts, João Gonçalves, Jovan Golić, Jukka Mäkelä, Karl Palmskog, Kostas Pentikousis, Lars Voelker, Laurent Mathy, Leonard Pitzu, Liang Zhao, M. Ángeles Callejo Rodríguez, Mads Dam, Marco Marchisio, Marcus Brunner, Mario Kind, Matteo D'Ambrosio, Melinda Barabas, Michael Kleis, Miguel Ponce de Leon, Mohammed Achemlal, Olli Mämmelä, Ove Strandberg, Panagiotis Papadimitriou, Patrick Phelan, Rebecca Steinert, René Rembarz, Reuven Cohen, Rolf Stadler, Rui Aguiar, Sabine Randriamasy, Teemu Rautio, Thomas Monath, Thomas-Rolf Banniza, Vinicio Vercellone, Virgil Dobrota, Yasir Zaki, Zakaria Khan, Zsolt Polgar, Zsuzsanna Kiss

List of Acronyms

3G	Third Generation
3GPP	3rd Generation Partnership Project
4G	Fourth Generation
AAA	Administration, Authorization, and Authentication
ACK	Acknowledgment
AdHC	Ad-Hoc Communities
AHDR	Ad-Hoc Disaster Recovery
AM	Anchorless Mobility
AN	Access Node
AODV	Ad-Hoc On-Demand Vector
AP	Access Point
API	Application Programming Interface
ARP	Address Resolution Protocol
ARQ	Automatic Repeat Request
AS	Autonomous System
ASN	Autonomous System Number
BE	Best Effort
BER	Bit Error Rate
BEREC	Body of European Regulators
BFD	Bidirectional Forwarding Detection
BGP	Border Gateway Protocol
BIOS	Basic Input/Output System
BLER	Block Error Rate
BO	Bit-level Objects
BU	Binding Update
CA	Channel Assignment
CAIDA	Cooperative Association for Internet Data Analysis
CAPEX	Capital Expenditure
CBA	Component Based Architecture
CBR	Constant Bit Rate
CBSE	Component Based Software Engineering

CCFW	Cooperation and Coding Framework
CCN	Content Centric Networks
CEP	Connected End Points
CF	Cooperation/Coding Facility
CFL	CF Layer
CLQ	Cross-Layer QoS
CMT	Concurrent Multipath Transfer
CN	Correspondent Node
Co-AD	Content Adaptation
CPU	Central Processing Unit
CRC	Cyclic Redundancy Check
CSMA	Carrier Sense Multiple Access
CSMA/CD	Carrier Sense Multiple Access/Collision Detection
CT	Compartment
CTR	Compartment Record
DA	Deviation Advertisement
DBA	Dynamic Bandwidth Allocation
DCF	Dispersion Compensating Fiber
DDOS	Distributed Denial of Service
DF	Digital Fountain
DGE	Dynamic Gain Equalizers
DHCP	Dynamic Host Configuration Protocol
DHT	Distributed Hash Table
DIF	Distributed IP Facility
DL	Downlink
DMA	Dynamic Mobility Anchoring
DMV2	Data-Multimedia-Voice-Video
DNC	Deterministic Network Coding
DNS	Domain Name System
DONA	Data Oriented Network Architecture
DOS	Denial of Service
DSL	Domain Specific Language, or Digital Subscriber Line
DTN	Delay/Disruption Tolerant Network
E2E	End-to-End
EC	European Commission
ECN	Explicit Congestion Notification
EFCP	Error and Flow Control Protocol
EGP	Exterior Gateway Protocol
EMT	Emergency Medical Team
EP	End Point
EPON	Ethernet Passive Optical Network
ERC	Emergency Response Command
ERG	European Regulators Group
ETT	Expected Transmission Time
ETX	Expected Transmission Count

FARA	Forward Directive, Association, and Rendezvous Architecture
FB	Functional Block
FCAPS	Fault, Configuration, Accounting, Performance, and Security
FDP	Forwarding Decision Process
FEC	Forward Error Correction
FER	Frame Error Rate
FI	Future Internet
FIA	Future Internet Architectures
FIB	Forwarding Information Base
FIFO	First-In First-Out discipline
FIM	Flow Interception Module
FIND	The future Internet design
FL	Folding Link
Fl-EP	Flow Endpoint
Fl-RO	Flow Routing
FN	Folding Node
FNE	Forwarding NE
FO	Fixed Operator
ForCES	Forwarding/Control Element Separation
FP7	Framework Programme 7
FPNE	Flow Processing NE
FQ	Fair Queuing
FRR	Fast Reroute
FSA	Flow State Advertisement
FTP	File Transfer Protocol
FTTH	Fiber to the Home
GAP	Generic Aggregation Protocol
GEF	Graphical Editing Framework
GENI	Global Environment for Network Innovation
GF	Galois Field
GGAP	Gossip-Generic Aggregation Protocol
GMOPR	Grid MOPR
GMP	Global Management Point
GMPLS	Generalized Multi-Protocol Label Switching
GMPR	Generic Path Master Record
GP	Generic Path
GPMR	Generic Path Management Record
GPRS	General Packet Radio Service
GPS	Global Positioning System
GRDF	Generic Resource Description Framework
GRX	GPRS Roaming Exchange
GRX	GSM Roaming Exchange
GSM	Global System for Mobile Communications
GSMA	GSM Association
GS-Node	Governance Stratum Node

GUI	Graphical User Interface
HA	Home Agent
HEN	Heterogeneous Experimental Network
HIP	Host Identity Protocol
HTTP	Hypertext Transfer Protocol
iAWARE	Interference Aware routing metric
ICANN	Internet Corporation for Assigned Names and Numbers
ICN	Information-Centric Network
ICVNet	Interconnecting Virtual Network
ID	Identifier
IDR	Inter Domain Routing
IETF	Internet Engineering Task Force
IGP	Interior Gateway Protocol
ILA	Interference-Load Aware routing metric
ILC	Inter Layer Communication
ILR	Inter Layer Routing
ILS	Information Lookup Service
INI	Information Network Interface
INM	In-Network Management
InP	Infrastructure Provider
IO	Information Object
IOLS	Information Object Lookup Service
IP	Internet Protocol
IPC	Inter-Process Communication
IPTV	Internet Protocol Television
IPv6	Internet Protocol version 6
IPX	IP packet eXchange
ISP	Internet Service Provider
IT	Information Technologies
ITU	International Telecommunication Union
IXP	Internet Exchange Point
JSIM	JavaSim
KS-Node	Knowledge Stratum Node
LAN	Local Area Network
LLC	Late Locator Construction
LLID	Logical Link ID
LQO	Link Quality Ordering
LQODV	Link Quality Ordering-based Distance Vector
LSA	Link State Advertisement
LSP	Label Switched Path
LSR	Label Switch Router
LT	Luby Transform
LTE	Long Term Evolution
MAC	Media Access Control
MANET	Mobile Ad Hoc Network

MAP	Mesh Access Point
MBMS	Multimedia Broadcast/Multicast Service
MC	Management Capabilities
MDHT	Multiple Distributed Hash Table
MED	Multi-Exit Discriminator
MEE-GP	Multihomed End-to-End GP
MEEM	Multihomed End-to-End Mobility
MIC	Metric of Interference and Channel-switching
MIH	Media Independent Handover
MILP	Mixed Integer Linear Program
MIP	Mobile IP
MMS	Multimedia Messaging System
MN	Mobile Node
MNE	Mediating NE
Mo-AH	Mobility Anchor
MOPR	Multi-Objective MPR
MP	Mediation Point
MP2MP	Multipoint-To-Multi-Point
MP2P	Multipoint-To-Point
MP-BGP	Multi-Protocol BGP
MPC	Multi-Party Computation
MPLS	Multiprotocol Label Switching
MPR	Multi-Path Routing
MPR-CT	MPR Compartment
MPR-GP	MPR GP
MPR-ME	MPR Master Entity
MR	Master Record
MS-Node	Machine Stratum Node
MTU	Maximum Transfer Unit
NACK	Negative Acknowledgment
NAT	Network Address Translation
NATO!	Not all at once
NC	Network Coding
NDL	Network Description Language
NE	Networking Entity
NED	NEtwork Description language (of OMNeT++)
NetInf	Network of Information
NGN	New Generation Network
NHLFE	Next Hop Label Forwarding Entry
Ni-IO	NetInf Information Object
Ni-MG	NetInfo Manager
NIN	NetInf node
NLRI	Network Layer Reachable Information
Node	CT Node Compartment
NR	Name Resolution

NRS	Name Resolution Service
NSF	National Science Foundation
NSIS	Next Steps in Signalling
NSLP	NSIS Signalling Layer Protocol
NTP	Network Time Protocol
NW	Network
OADM	Optical Add Drop Multiplexers
OCS	Optical Circuit Switching
OD	Origin Destination
OFDM	Orthogonal Frequency Division Multiplex
OLA	Optical Line Amplifiers
OLT	Optical Line Terminal
OM	Observation Module
ONU	Optical Network Unit
OPEX	Operational Expenditure
OS	Operating System
OSGi	Open Services Gateway initiative
OSI-SM	Open Systems Interconnection—System Management
OSPF	Open Shortest Path First
OSPF-TE	Open Shortest Path First—Traffic Engineering Extensions
OSS	Open-Source Software
OTN	Optical Transport Network
OWL	Web Ontology Language
OXC	Optical Cross-Connect
P2MP	Point-to-Multipoint
P2P	Peer to Peer
Pa-EP	Path Endpoint
Pa-RO	Path Routing
PC	Personal Computer
PCE	Path Computation Element
(P)CN	Congestion and Pre-Congestion Notification
PDU	Protocol Data Unit
PER	Packet Error Rate
PHY	Physical Layer
PIM-DM	Protocol Independent Multicast—Dense Mode
PIM-SM	Protocol Independent Multicast—Sparse Mode
PMD	Polarization Mode Dispersion
PMS	Personal Mobile Scenario
PN	Provisioning Network
PnP	Plug-and-Play
PNP	Physical Network Provider
Po-EN	Policy Engine
PON	Passive Optical Network
PPP	Point-to-Point
PSR	Packet Success Rate

PSTN	Public Switched Telephone Network
QoE	Quality of Experience
QoS	Quality of Service
R&D	Research and Development
RA	Resource Advertisement
RAID	Redundant Array of Inexpensive Disks
RDF	Resource Description Framework
RDL	Resource Description Language
RDP	Routing Decision Process
RESCT	Resolution Compartment
RFC	Request For Comments
RFID	Radiofrequency Identification
RI	Routing Instruction
RIB	Routing Information Base
R-MAC	Radio Medium Access Control
RNC	Random Network Coding
RNE	Routing NE
RNG	Random Number Generator
RO	Routing Object
ROADM	Reconfigurable Optical Add-Drop Multiplexer
RSVP	Resource Reservation Protocol
RSVP-TE	Resource Reservation Protocol—Traffic Engineering
RTT	Round-Trip Time
RUI	Routing Update Interval
SA	Service Agent
SAP	Service Access Point
SATO	Service-Aware Transport Overlay
SCTP	Stream Control Transmission Protocol
SDH	Synchronous Digital Hierarchy
SE	Self-managing Entities
SGP	Service Gateway Point, or Stratum Gateway Point
SHIM6	Site Multihoming by IPv6 Intermediation
SIM	Subscriber Identity Module
SINR	Signal to Interference and Noise Ratio
SIP	Session Initiation Protocol
SLA	Service Level Agreement
Sl-MA	Service Level Agreement Manager
SNMP	Simple Network Management Protocol
SNR	Signal to Noise Ratio
SOA	Service Oriented Architecture
SOCP	Second Order Conic Program
SON	Service Oriented Networks
SONET	Synchronous Optical Network
SP	Service Provider
SQF	Shortest Queue First discipline

SRDF	Semantic Resource Description Framework
SRMF	Semantic Resource Management Framework
sRTT	smoothed Round-Trip Time
SSDP	Simple Service Discovery Protocol
SSP	Service Stratum Point, or Stratum Service Point
SVN	SubVersioN
SW	Software
Tagg-GP	Transport aggregate GP
TCG	Trusted Computing Group
TCP	Transmission Control Protocol
TDM	Time Division Multiplex
TDMA	Time Division Multiple Access
TE	Traffic Engineering
TENE	Traffic Engineering NE
TGP	Transport GP
TIC	Time Interval Counter
TLS	Transport Layer Security
TM	Transformation Module
TMN	Telecommunications Management Network
TO	Time-out
To-DB	Topology Database
Tr-MO	Traffic Monitoring
TTL	Time to Live
TTM	Time To Market
TV	Television
UA	User Agent
UDP	User Datagram Protocol
UIP	Unmanaged Internet Protocol
UL	Uplink
UML	Unified Modelling Language
UMTS	Universal Mobile Telecommunications System
UPnP	Universal Plug and Play
URL	Uniform Resource Locator
VBR	Variable Bit Rate
Vi-Node	Virtual Node
VLAN	Virtual Local Area Network
VLC	VLC media player
VNet	Virtual network
VNM	VNet Management
VNO	VNet Operator
VNP	VNet Provider
VoD	Video on Demand
VoIP	Voice over IP
VPN	Virtual Private Network
WAN	Wide Area Network

WAP	Wireless Access Point
WCETT	Weighted Cumulative ETT
WDM	Wavelength Division Multiplexing
WFQ	Weighted Fair Queuing discipline
WiFi	Wireless Fidelity
WiMAX	Worldwide Interoperability for Microwave Access
WLAN	Wireless Local Area Network
W-LLC	Wireless Link Layer Control
WMAN	Wireless Metropolitan Area Networks
WMN	Wireless Mesh Network
WMOPR	Wireless MOPR
WMVF	Wireless Medium Virtualization Framework
WP	Work Package
xDSL	Digital Subscriber Line
XML	eXtensible Markup Language

List of Figures

List of Tables

Chapter 1
Introduction

**Luis M. Correia, Henrik Abramowicz,
Martin Johnsson, and Klaus Wünstel**

Abstract It starts by addressing some of the problems with current Internet networks, its core architecture and its evolution model. Current architectures are typically developed around layered models, and deficiencies have been shown, such as lack of support for QoS and seamless mobility, security vulnerabilities, and address shortage. The various forms of unwanted traffic, including spam, distributed denial of service, and phishing, are arguably some of the biggest problems. Changing business models are likely to have impact on the network. Privacy and accountability are other major issues. Next, a brief description of the 4WARD project is provided. It is followed by an overview of the current research and development activities being held in Europe, USA and Japan. It ends with a global view of the contents of the book.

1.1 Problems with Current Internet

The discussion on the "Network of the Future" is gaining in intensity due to increasing concerns about the inability of the current Internet to address a number of important issues affecting present and future services and to the impetus provided by "clean slate design" research initiatives launched in the US, Europe and Asia.

L.M. Correia (✉)
IST/IT—Technical University of Lisbon, Lisbon, Portugal
e-mail: luis.correia@lx.it.pt

H. Abramowicz · M. Johnsson
Ericsson Research, Stockholm, Sweden

K. Wünstel
Alcatel Lucent Bell Labs, Stuttgart, Germany

L.M. Correia et al. (eds.), *Architecture and Design for the Future Internet*,
Signals and Communication Technology,
DOI 10.1007/978-90-481-9346-2_1, © Springer Science+Business Media B.V. 2011

Many problems with the current network architecture have been recognised for a long time, but have not received a satisfactory solution. Below is a description of a number of issues pertaining to the current Internet. It is just example of problems, and not an exhaustive list of issues.

It should be remembered that the current Internet was initially developed for a world in which a limited number of trusted nodes interconnected by copper based transmission technology implemented distributed applications mostly some kind of file transfer and message exchange. The initial architecture developed for this purpose was essentially simple, but open for new applications. Its evolution has led to a tremendous success—the Internet as we know it today. It is however far from clear that it is still the optimally evolvable solution, able to meet the challenges of dominating fibre optics and radio transmission technology, real-time multimedia and file-sharing applications and exposure to an untrustworthy world. Furthermore the Internet, starting as a simple set of protocols and rules, has over the decades reached a state of high complexity with regard to interoperability, configuration and management.

1.1.1 Internet's Core Architecture and Evolution Model Is No Longer Suitable

The predecessor of today's Internet, the Arpanet, started in the late 1960s as a network of four university networks. It was created as a fixed network for hosts that were neither mobile nor wirelessly connected. Today, firewalls, network address and port translators, as well as session border controllers decouple the different IP networks at various layers. The capability of **all** end nodes to act as both consumer and producer of information has been continually reduced and, in the case of mobile nodes, may be considered not to exist at all; IPv6 failed to provide the necessary general architectural overhaul the Internet needs to become the universal Network of the Future that people can rely on. Worse, there are good reasons to believe that a single network solution will not be able to cover and satisfy the future needs in networking.

While the scale of the Internet has not yet reached its limit, the growth in functionality, i.e., the ability of the global system to adapt to new functional requirements, has almost come to a standstill. We have reached a critical point in an impressive development cycle that now requires a major change. 4WARD is taking a long term perspective on network (r)evolution and adopts a clean-slate research approach. We expect this research effort to have major industrial impact during the next decade.

1.1.2 Ossification of Internet

Current architectures for communication systems are typically developed around layered models (e.g., Internet, OSI, 3GPP). Practice in the open Internet environment has shown that it is difficult to deploy network enhancements, such as IPv6, IPSec, MobileIP, or multicast. The dramatic growth of the Internet has also brought into sharp focus its architectural deficiencies such as lack of support for QoS and seamless mobility, security vulnerabilities, address shortage, to name a few. Although a number of solutions have been proposed for these problems, these can, at best, be described as a patchwork of fixes to fill architectural holes. Furthermore, most of them have come about in an uncoordinated and ad-hoc fashion and hence, they have spawned problems of their own.

The resulting system has become quite complex, often with similar functionality re-appearing over and over again in different protocols and layers. The piecemeal ad hoc approach to solving problems that applies patches to certain parts of the protocol stack can in fact jeopardise the operation and performance of other parts of the communication system.

1.1.3 Surge of Unwanted Traffic, Including but Not Limited to SPAM

The various forms of unwanted traffic, including spam, distributed denial of service (DDoS), and phishing, are arguably the biggest problem in the current Internet. Most of us receive our daily dosage of spam messages; the more lucky of us just a few of them, the more unlucky a few hundreds each day. Distributed denial of service attacks are an everyday problem to the large ISPs, with each major web site or content provider getting their share. And, as we all know, phishing is getting increasingly common and cunningly sophisticated.

The root reasons to unwanted traffic seem to be best characterised with economics. We can characterise the current Internet as a global, distributed message passing system where the recipient pays the main cost of unwanted communication. This is a direct (though certainly unintentional) consequence of the network architecture. By explicitly and directly naming all the potential recipients, we create a system where the senders can easily express their desire to send data to any recipient in the network. Given that, under the typical flat-fee contracts, the marginal cost of sending additional packets is very close to zero (up to some capacity limit). Hence, there are few or no incentives for refraining from sending unwanted traffic; sending some more packets, either just for fun in order to gain legitimate or illegitimate profits, costs so little that it doesn't matter. Hence, for SPAM, even a marginal response rate creates a strong incentive for sending unsolicited advertisements, and for DDoS-based extortion, even a small success rate creates a strong incentive to launch attacks.

To summarise the current unwanted traffic problem is a compound result from the following factors:

- An architectural approach, where each recipient has an explicit name and where each potential sender can send packets to any recipient without the recipient's consent.

- A business structure, where the marginal cost of sending some more packets or messages (up to some usually quite high limit) is very close to zero.
- The lack of laws, international treaties, and especially enforcement structures that would allow effective punishment of those engaging in illegal activity in the Internet.

Basically, the separation of identifiers and locators can be used to create architectures where a sender must acquire the recipients' consent before it can send any data, beyond some severely rate-limited signalling messages.

1.1.4 Configuration and Management Complexity

Networks are becoming larger, more heterogeneous, and more dynamic. End users expect ubiquitous service availability on a variety of devices and equipment. More equipment and network types will coexist in a single network operator's domain. Security threats will change as networks and services change. Changing business models are likely to require that network elements are able to enforce access control locally and for instance maintain configuration integrity despite allowing access to important resources from different administrative domains.

The traditional operator-to-subscriber based business models are being replaced by several other models, such as the user-to-network model. The emerging user-to-user model is a challenging model for operators: Web 2.0 services with user-generated content exist only in the service plane. Typically, end-user services are composite Data-Multimedia-Voice-Video (DMV2) services. In this model, users get access to the content of other users, with revenue generation typically provided by advertising revenue. Outsourcing of management entirely or partially to one or more outsourcing providers is another common business relationship requiring means for tracking service delivery liability. As a consequence of this business innovation, the technical network operations perspective of the traditional operator is being substituted with a business-focused service management perspective, where service delivery according to end-user expectations is crucial and lower level network aspects are not significantly interesting.

The total cost of ownership for service enabling equipment is highly focused by operators. At present, there are typically numerous heterogeneous management displays from different vendors. They do not provide sufficient input to business decisions and prioritisations, which makes deployment and assurance of even a very small service a time-consuming challenge. Being to a large extent based on humans, this management doesn't scale.

From a vendor perspective, the current ad-hoc design of element and network management instrumentation and systems is costly in relation to business value for the service provider. Network element instrumentation consist of hundreds of performance counters, events, alarms and configuration parameters. This challenges both network element vendor management system developers and the operations staff.

1.1.5 Lack of Privacy and Accountability

The aim of privacy and accountability is to prevent socially undesirable things from happening, on one hand by imposing technical restrictions on information flow, and on the other hand by creating explicit incentives for desirable behaviour.

The privacy problem is a complex one, with at least three different viewpoints. From the Orwellian point of view, the question is about freedom of speech and governmental control. A sufficient privacy system ensures that we can express our opinions and think freely, within reasonable bounds (like not committing clearly criminal acts) even when our opinions are socially unacceptable or hostile towards the governing regime. The Kafkaesque aspect of privacy focuses on citizen's ability to retain their autonomy without fear of unfounded litigation or other harassing legal/other action. Thirdly, the economic aspect of privacy relates to the fine balance between socially beneficial differentiated pricing vs. socially harmful price discrimination. From these three different points of view, it seems a necessity to provide a reasonable base-level of privacy as a built-in feature in future networks.

The flip side of privacy is accountability. Unbounded privacy encourages irresponsible behaviour patterns, such as rampant advertising. To counter these, increased privacy requires increased accountability; a fact that appears as a paradox from the technical point of view. A key to understanding this technical paradox is to consider the different dimensions of communication. At the baseline level, we can make a difference between four dimensions: the content of communication, the parties communicating, their locations, and finally the very fact that a piece of communication took place (existence). If the system is able to provide strong "insulation" between these dimensions so that each party gets only the relevant pieces of information, a high level of privacy can be preserved. For example, a communications service provider needs to know that communication takes place and whom to attribute the communication to, but should have no access the content, the identity of the other parties, nor their locations.

1.1.6 Poor Support for Mobility and Multi-homing

Effective mobility support requires a level of indirection. It is needed to map the mobile entity's stable name to its dynamic, changing location. Effective multi-homing support (or support for multi-access/multi-presence) requires a similar kind of indirection, allowing the unique name of a multi-accessible entity to be mapped to the multitude of locations at which it is reachable.

Within the Internet community, the classical approach has been to consider mobility and multi-homing as separate, technical problems. The main result of this are the Mobile IP protocols, which are architecturally based on re-using a single name space, the IP address space, for both stable host identifiers (Home Addresses) and dynamic locators (Care-of Addresses). While the approach certainly works, it creates two major drawbacks. At the same time, the tendency of considering multi-homing a separate problem with a separate solution creates feature interactions.

With regard to the Mobile IP approach, it binds the communication sessions (TCP connections and application state) to the home addresses. This, in turn, when combined with the only known scalable solutions to a number of related security problems, creates an undesirable dependency on a constant reachability of the home address. In other words, the Mobile IP architecture is intrinsically bound to the availability of the home addresses; the home agent becomes a new single point of failure.

Secondly, approaches that use names from a single name space for multiple purposes create a number of potential semantic problems. When Mobile IP is used, there are no easy way to tell if two IP addresses actually point to a single host (e.g., due to one being its home address and another one its care-of address) or not, i.e., whether one is merely an alias for the other or an identifier for a genuinely different node. That, in turn, may lead to very confusing problems for quite a large number of applications.

1.2 Short 4WARD Overview

4WARD performs research on the architecture of a Future Internet adopting a "clean slate" research approach. This means the practical constraint of evolving from the existing TCP/IP-based network architecture is temporarily ignored in the interest of discovering a design that is ideally adapted to present and expected future usage and is not forced to adapt to architectural decisions made some thirty years ago with quite different objectives and constraints. An architecture designed following this approach may be seen as a target for the current network to evolve to. It may alternatively be seen as the blueprint of a parallel architecture that could coexist and interoperate with IP, gradually expanding and taking over the functions of the old network.

The strategic objective of 4WARD is to increase the competitiveness of the European networking industry and to improve the quality of life for European citizens by creating a family of dependable and interoperable networks providing direct and ubiquitous access to information. 4WARD's goal is to make the development of networks and networked applications faster and easier, leading to both more advanced and more affordable communication services.

To achieve this strategic objective, work in 4WARD is guided by 4 overriding tenets:

- **Tenet 1: Let 1000 Networks Bloom**
  The project explores a new approach to the creation and co-existence of a multitude of networks: the best network for each task, each device, each customer, and each technology. 4WARD aims to create a framework in which it will be easy for

many networks to bloom as part of a family of interoperable networks that can co-exist and complement each other.

- **Tenet 2: Let Networks Manage Themselves**
  The 4WARD architecture incorporates an embedded management entity, which is an inseparable part of the network and each of its components, generating extra value in terms of guaranteed performance in a cost effective way, and capable of adjusting itself to different network sizes, configurations, and external conditions under the control of policies set by the network owner.
- **Tenet 3: Let a Network Path Be an Active Unit**
  A forwarding path is recognised as an active network component that controls itself and provides customised transport services. An active path can provide resilience and fail-over, offer mobility, simultaneously use multiple different sequences of links, secure and compress transmitted data, and optimise its performance.
- **Tenet 4: Let Networks Be Information-Centric**
  Users are primarily interested in using services and accessing information, not in the nodes that host information or provide services. Consequently, the 4WARD architecture considers information objects (and their digital instantiations) and services as primary importance that are not tied to any particular device but can rather be mobile and distributed throughout the network. Such, 4WARD addresses one of the fundamental flaws of the Internet architecture.

The Future Internet will be even more important for society at large than the present network and 4WARD therefore also performs research on the socio-economic and regulatory issues arising from the application of the above tenets.

In our approach, we combine on the one hand innovations needed to improve specific aspects of a network architecture, and on the other hand work on a common overall architecture framework that neatly fit these innovations together.

This work is structured into six work packages: three of them consider innovations for a single network architecture, i.e., Generic Path, In-Network Management and the Network of Information, one work package studies the use of Virtualisation to allow multiple networking architectures to co-exist on the same infrastructure, another work package looks at the design and development of Interoperable Architectures, and finally one work package that ensures that all envisaged developments take proper account of essential Non-Technical Issues.

4WARD is an Integrated Project assembling 36 partners in a strong, industry-led consortium of the leading operators, vendors, SMEs, and research organisations. The consortium includes partners from North America and Asia and has a strong background of research on networking architecture with particular expertise in the field of wireless and mobility. The project has originally been granted a budget of 23 M€ for a period of two years, but it has been extended for another half year to match with future Call 5 projects.

1.3 Position of 4WARD in Europe and EC Projects and Other Regions

1.3.1 EU Framework Programme 7

In the EU Framework Programme 7 (FP7) there are several projects that are relevant for this area. Some of them are having a clean slate approach others are working with an incremental approach to resolve some of the issues in the current Internet within the present paradigm.

A couple projects that are relevant for this area are listed below:

- PSIRP is a STREP with the aim to investigate the "publish–subscribe" paradigm.
- Trilogy is an IP lead by BT to resolve the current problems with BGP within the present paradigm.
- Sensei is an IP working with sensor networks and trying to create open *service interfaces* and corresponding semantic specification to unify the access to context information and actuation services offered by the system for services and applications.
- Onelab2 is an IP that deals with creation of a testbed for experiments for Future Internet.
- Moment is a STREP dedicated to handle bandwidth measurement.

The Commission is, in addition, trying to coordinate the Future Internet activities more actively and has also established a Future Internet Assembly amongst the research projects to further the activities on Future Internet and coordinate across a number of domains like content media, security networking, etc. 4WARD has played a prominent role in FIA and have responsibilities as caretakers helping to organise the meetings and sessions,

4WARD has through its coordinator also been driving the Future Internet clusters where a lot of architecture and scenario work has taken place and been used as input also to the Future Internet Assembly.

1.3.2 FIND (Future Internet Design) US

The Future Internet Design (FIND) program was initiated by National Science Foundation (NSF) in 2006 with the objective of supporting a wide range of small-to-medium sized "clean-slate" protocol investigations across the academic research community. The scope of the program includes trust, security, impact of emerging wireless and optical technologies, network economics and social aspects.

In 2009, NSF organised an external panel review of the FIND program, involving a detailed evaluation of over 30 projects. The panel provided a strong positive recommendation about the program, commenting on the benefits of clean-slate research without the usual constraints of backward compatibility with existing network protocols. The panellists felt that new ground was being broken on important research topics such as: naming, addressing, routing, monitoring, mobility, network management, access and transport technologies, sensing, content and media delivery, and networked applications. The panel recommended that NSF continue the program and initiate an integrated community effort to build teams who would design and prototype more comprehensive converged future Internet architectures. The panel also recommended an increased focus on security and network management aspects. NSF accepted these conclusions and formed a new program called "Future Internet Architectures (FIA)" (NSF 10-528) that would support 2–4 large project teams working on comprehensive and converged future Internet architectures. These projects are expected to result in a completed design, protocol validation and initial deployment on infrastructures such as GENI.

1.3.3 GENI (Global Environment for Network Innovation) US

The GENI (Global Environment for Network Innovation) program was initiated by NSF in 2008 with the objective of developing flexible and large-scale networking infrastructure for future Internet research being done under FIND (Future Internet Design) and other programs. GENI is managed by the GPO (GENI project office) at BBN Technologies, Cambridge, MA and is headed by Chip Elliott, Program Manager. The approach adopted by GENI is based on a number of principles including:

- Spiral development with continuous improvement and feedback
- Leveraging existing capabilities and testbeds across US research community
- Federation of testbeds and campus networks to form an integrated GENI facility
- Competition among research groups for selection of key GENI components
- Open, collaborative project with open-source software, international partners, etc.

GENI has been organised into Spirals, with Spiral I starting in Nov 2008 and ending in Nov 2009, and Spiral II starting in Dec 2009 and ending in Dec 2010. The first spiral emphasised technology evaluation and risk mitigation through proof-of-concept prototypes. The second emphasises integration of an initial federated "meso-scale" GENI prototype across ~8–10 campus locations, with a unified experimental control and management interface.

1.3.4 Akari Japan

The objective of the AKARI Architecture Design Project (in short AKARI Project) is to design the network of the future. The AKARI Project aims to implement a new generation network by 2015 by establishing a network architecture and creating a network design based on that architecture. The motto is "a small light (*akari* in Japanese) in the dark pointing to the future." The philosophy is to pursue an ideal solution by researching new network architectures from a clean slate, without being impeded by existing constraints. Then the issue of migration from existing networks can be considered. The goal is to create an overarching design of what the entire future network should be. To accomplish this vision of a future network embedded as part of societal infrastructure, each fundamental technology or sub-architecture must be selected and the overall design simplified through integration.

The AKARI project schedule is divided into two five-year periods: the first five-year period (FY 2006–2010) aims at finalising the new generation network design blueprint and the second five-year period (FY 2011–2015) will develop test-beds based on the blueprint. In the first year (FY 2006), the conceptual design was created and initial design principles were presented. Detailed design was performed during the second year while revising the initial design principles. Prototypes will be developed, evaluated, and verified to indicate the validity of the concepts. Design diagrams will be completed in the fifth year in the first five year period.

In the sixth and subsequent years, the new generation network concepts will be incorporated in test-beds based on the developed prototypes and design diagrams to conduct demonstration experiments. In addition, the network components will be created and protocol engineering will be performed to establish new generation network construction techniques.

1.4 The Book

The book from the 4WARD project deals with Architecture and Design for the Future Internet and is covering a broad spectrum of issues. We give a system overview and give a socio-economic background reasons and regulations for a Future Internet, but also go into some depth of the different technical issues.

Chapter 2 on System Overview describes the System Model and defines the structure and behaviour of a system that is to be constructed as well as its generativity, i.e., how bigger and more complex future systems and networks could be built by using a small set of generic concepts. 4WARD is promoting a new approach to networking based on the analysis of both the success factors of the Internet (seen as the core Internet design principles and core IP protocols) as well as the factors that have led to ossification and the patchwork type of the IP evolution of recent years. The Network of the Future must be based on a new set of architectural principles.

It is well understood that the development path of any industry or economic sector is significantly affected by the opportunities provided by the available technologies, the particular characteristics of its markets and the directions and priorities of related government policies and regulations. Previously, there has been a tendency to leave these issues to be handled separately, and the non-technical topics above have been addressed after the technology had been developed. In the case of the global networked society, this is not a desirable approach. The take-off and success of the Future Internet will be closely linked with what actions are taken on all areas of the Future Internet ecosystem. Chapter 3 of the book on Socio-economic aspects describes how major non-technical drivers impact the transition from the

R&D stage to the real deployment of the technical and architectural innovations studied in 4WARD.

Virtualisation is a key technology for the deployment of new customised network architectures. After a short introduction into the overall concept of network virtualisation, its goals and benefits as well as scenarios and business aspects are presented. Then the virtualisation framework is described in more detail, starting with an overview of the process for building and setting up virtual networks, and including resource virtualisation, and provisioning, control and management of virtual networks. Afterwards the design process to be followed by the network architect, the design of new network architectures is described. In this process, the network architect can also follow the design patterns described in Chap. 4 on "How to design and build networks" in order to: (i) effectively compose different functionalities to meet the initial requirements, and (ii) assure the interoperability among different architectures, taking business relationships, security and management issues into account. In order to assure the interoperability among virtual networks, the concept of folding points is analysed in detail.

Naming and addressing has been a source of considerable contention in existing network designs. What precisely is named, what an address is, and how these two concepts relate to each other by name resolution has been treated differently and inconsistently in different systems as well as in different architectures. The 4WARD project pursues an integrated, coherent approach for a naming & addressing architecture that combines flexibility with coherence and integrates its different components via a cross-layer name resolution concept. Chapter 5 on Naming and Addressing will discuss the basic design rationale of this concept. It will also go over some examples, ranging from very simple, local naming/addressing schemes, over schemes intended for consistent naming and addressing in a network layer extending world-wide, to a rather complex naming/addressing structure suitable for a data-centric network of information. All these schemes combine into the overall naming and name resolution architecture, yet remain flexible at their respective layers of abstraction.

Chapter 6 on Security principles gives some considerations for how rethinking the fundamental network architecture affects and is driven by security considerations. The information-centric approach of 4WARD is built on the concept of securing information rather than the containers containing information. Doing so, the security principles based on ownership and controlling access at the originating source become challenged. At the same time, moving intelligence into the network itself challenges the underlying assumption of having an Internet consisting of neutral, dumb, and fundamentally cooperating and trusting autonomous domains. 4WARD has only begun addressing the security principles necessary for dynamical management of virtualised, largely self-configuring entities having specific properties. The specific security implementation choices necessary for network design, transport, routing, lookup, privacy, accountability, caching and monitoring are largely out of scope. 4WARD acknowledges and considers the business and governmental control interests that will heavily influence the security direction into which the future network evolves.

One of the key challenges for the Future Internet is the correct definition and implementation of the domain concept. Chapter 7 analyses the interconnection model of the Internet and of current Mobile Operators and presents the inter-domain concepts developed in the scope of 4WARD. Special attention is devoted to the still to solve problem of Multi-domain Quality of Service.

The cost and complexity of configuring and running networked services are significant and expected to increase. We propose a solution for management, *In-Network Management* (INM), which is based on decentralisation, self-organisation, and autonomy of management processes. Its key idea is that management stations outside the network delegate management tasks to the network itself, supporting future large-scale networks that self-configure, dynamically adapt to external events and allow for low-cost operation. In this Chap. 8 on "How to manage networks", we will discuss challenges, benefits, and approaches to In-Network Management. We present an architectural framework suitable for different levels of embedding within the network elements. Examples of novel algorithms supporting real-time monitoring in a distributed manner are presented, and self-adaptation schemes for resource control are discussed.

Transporting information through the Internet has traditionally been following the end-to-end principle. This means that no knowledge about the nature of the transported information is assumed within the network and leads consequently to overlay networks realising specific services. Keeping state information "in the network" is generally seen as a burden for scalability and undesirable. However, mobility of hosts and applications, any guarantees for quality of service, and new methods for cooperation and coding, all require a certain amount of information to be stored at specific places inside the network. Chapter 9 describes an architecture for data transmission that puts technological and administrative domains (compartments) in the role of the keeper of this shared information. Paths are established between communicating entities, basic functional blocks that re-appear in different layers of the Internet. We explain how certain functions like routing, access control, and resource management are recurring in entities at all layers, and therefore allow an object oriented definition of entities and paths. Compartments and generic paths limit the scope within which state information needs to be kept consistent. Compartment layering is fundamentally different from the established ISO/OSI model and the chapter discusses several examples for the use of cooperation between more than the traditional two end points of a transmission.

Chapter 10 presents the overall vision for a network of information, illustrates the fundamental ideas, and explains the mechanisms currently under development that will bring about a major paradigm change in networking. After briefly reviewing relevant scenarios where the current host-centric approach to information storage and retrieval is ill-suited for, we introduce how a new networking paradigm emerges, by adopting the information-centric network architecture approach. We illustrate how information retrieval may look like in the future, emphasising on the user perspective. We then put forward the architectural requirements for a network of information, and the research directions taken during the project. The core of this chapter centres on a lucid description of the mechanisms, the "nuts and bolts" so to speak, of the technologies that implement a network of information. We describe a network of information operation, providing concrete examples and highlighting the performance improvement expected to materialise with the deployment of a network of information. Finally, we take a long-term view and discuss how a network of information can evolve. This chapter concludes with a comprehensive summary of the main network of information innovations and future items of work.

In the preceding chapters, we have described concepts and technologies that can be used for designing and building networks, how networks can be interconnected and be managed, how connectivity can be established, and how to manage and search for information objects. In addition, we have also established important security principles and schemes for naming and addressing. Together, these provide a foundation and a set of tools for new ways of networking in the Future Internet. In order to show their advantages compared to current paradigms in networking, as well as to show how they can be applied in a consistent and coherent manner, Chap. 11 on Use Cases describes through a set of use cases how complete and integrated solutions for networking can be provided using the principles and tools described in chapters above. They will take us all the way from the design of suited network and software architectures, further on to describe how functionality and interfaces are being deployed, and finally how this functionality is being used and managed in order to carry out the specific tasks described by each of the use cases.

To support the theoretical ideas developed within 4WARD, some of them have been realised as prototypes. The experiences collected while implementing the different concepts gave valuable feedback and enhanced the ideas with crucial details. The most important concepts have been successfully tested, and this chapter will give an overview of the developed prototypes. Some of them are also publicly available. Pointers to the releases are given in the respective sections of Chap. 12 on Prototype Implementation.

The final Chap. 13 gives some conclusions and also describes some migration approaches by the 4WARD project to make it possible to realise the research findings.

Chapter 2
A System Overview

Martin Johnsson

Abstract The 4WARD System Model is described, defining the structure and behavior of a communication system that is to be constructed as well as its generativity, i.e., how bigger and more complex future systems and networks can be built
by using a small set of generic concepts. It presents the project four tenets. Then,
an Architecture Framework is shown, providing a unified component-based design
process, which defines a seamless step-wise though iterative process for deriving a
software-based network architecture using as input a set of technical requirements.
The Architecture Pillars, described in detail, are: In-Network Domain Management,
Network of Information, Generic Path, and the Physical Virtualized Substrate. The
Architecture Framework is presented in terms of Strata, Netlets, and the Design
Repository. The Design Process is also addressed.

2.1 Background and Motivation

This section describes the 4WARD System Model, which defines the structure and
behavior of a communication system that is to be constructed as well as its generativity, i.e., how bigger and more complex future systems and networks could be
built by using a small set of generic concepts.

Through 4WARD, a new approach to networking based on the analysis of both
the success factors of the Internet (seen as the core Internet design principles and
core IP protocols) as well as the factors that have led to ossification and the patchwork type of the IP evolution of recent years has been developed.

The Network of the Future must be based on a new set of *Internetworking principles*. These principles are characterized below as four programmatic tenets:

M. Johnsson (✉)
Ericsson Research, Stockholm, Sweden

L.M. Correia et al. (eds.), *Architecture and Design for the Future Internet*,
Signals and Communication Technology,
DOI 10.1007/978-90-481-9346-2_2, © Springer Science+Business Media B.V. 2011

1. Let 1000 Networks Bloom

   We will explore a new approach to a multitude of networks: the best network for each task, each device, each customer, and each technology. Unlike the multitude we had in the past, where different incompatible technologies were competing with each other, we want to create a framework that will allow these many networks to bloom as a family of interoperable networks coexisting and complementing each other.

2. Let Networks Manage Themselves

   The main limits of current technologies are the scaling up to very large network sizes, and the needed human intervention which is associated with considerable cost, errors and with an inherent slowness in reacting to changing network conditions. What we would like to have is a management entity as an inseparable part of the network itself, generating extra value in terms of guaranteed performance in a cost effective way, and capable of adjusting itself to different network sizes, configuration, and external conditions.

3. Let a Network Path Be an Active Unit

   We want to consider a path as an active part of the network that controls itself and provides customized transport services. An active path can provide resilience and fail-over, offer mobility, simultaneously use multiple different sequences of links, secure and compress transmitted data, and optimize its performance all by itself.

4. Let Networks Be Information-Centric

   Users are primarily interested in using services and accessing information, not in accessing nodes that host information or provide services. Consequently, we want to build a network as a network of information and services that may be mobile and distributed. In such a network, the users just accesses items of interest by their name while the data locations can be completely hidden.

These tenets, together with the understanding of the current situation of today's Internet, formed the main drivers for the definition of the 4WARD Technical Requirements [1], which laid the foundation for technical work within the 4WARD project. This work ultimately resulted in the 4WARD System Model, which is described in the following section.

2.2 The 4WARD System Model

Figure 2.1 depicts the 4WARD System Model, which has been developed with the Tenets and the 4WARD Technical Requirements [1] as main principal input. The system model gives the necessary definitions, specifications, principles and guidelines for designing, building, deploying, and manage interoperable network architectures. For that purpose, the 4WARD System Model consists of an Architecture Framework and a set of Architecture Pillars which provides the essential technologies in many of the network architectures anticipated and required for the future networks, though it is possible to also deploy and use them in migration scenarios. With the 4WARD System Model we expect significant efficiency gains in the de-

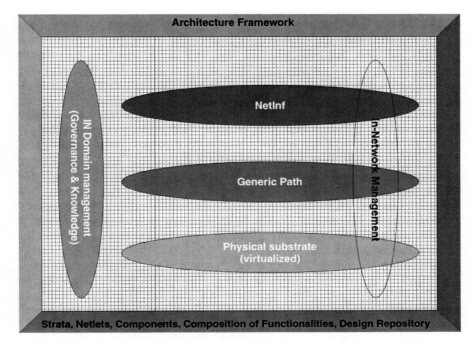

Fig. 2.1 The 4WARD System Model

sign, management and operation of networks, which is one of the key challenges in both current and future networks. The Architecture Pillars have been defined using a new set of concepts and technologies to address emerging business models and new types of applications:

- A new abstraction and model of the physical and virtualized infrastructure, including all of transmission, processing, and storage resources.
- ONE modular and extensible connectivity concept, supporting all modes and topologies of endpoint associations.
- A new open model and API for content and information management. Search and retrieval of information objects using a persistent identity.
- Management providing an inherent capability of the functions in the network.

The Architecture Framework provides a unified component-based design process, which defines a seamless step-wise though iterative process for deriving a software-based network architecture using as input a set of technical requirements. The design process includes the architectural principles and re-usable design patterns at various levels of abstractions out of which families of interoperable network architectures can be defined.

The Architecture Pillars: In-Network Domain Management, NetInf, Generic Path, and the Physical but virtualized substrate (and each of them in turn define their own respective frameworks or architectures) themselves to be defined by using the Architecture Framework.

The Physical Substrate provides an abstraction of the physical resources of any network spanning from the smallest to the largest. The abstraction is the key for coherent virtualization and management of those underlying resources across domain borders. The result of a virtualization operation is a virtualized network, providing resources onto which an operator is free to instantiate its own choice of functions, protocols, etc., for example Generic Paths and NetInf.

The Generic Path provides a generalized transport mechanism to transfer data between entities in the network. The recursive Generic Path concept is able to model virtually any type and level of transport, be it point-to-point or multipoint-to-multipoint, or supporting transport on links at the physical level, or end-to-end across networks. Generic Paths specifically give support for the dissemination of information objects.

NetInf (Network of Information) provides for identification, management, and dissemination of information objects. NetInf is a new abstraction of information (and service) management, where applications do not need to be aware of where an information object is stored.

In-Network Management (INM) is omni-present in all network functionalities. It provides design patterns and interfaces as well as more specific mechanisms, facilitating various degrees of self-management capabilities. This spans from such capabilities living 'beside' the functionality it is supposed to manage, and then all through to functionalities being fully and inherently self-managed.

A special case of In-Network Management is In-Network Domain Management, which provides self-management capabilities on domain as well as inter-domain scale. The Knowledge function (also known as Knowledge stratum) discovers, gathers and further infers status of network topologies, resource and context status by querying the network functionalities operating in the network, for example Generic Paths and NetInf. The Governance function (also known as Governance stratum) provides control and management of network functionalities, and governs by querying the status of the network from the Knowledge function. The Governance function will decide out from policies (provided by a network administrator) as well as the network status what network functionalities shall operate in the network. Governance and Knowledge functions are also instrumental for the interconnection and composition of networks and domains, where dynamic and highly automized creation of SLAs is supported.

The following sections provide an overview and introduction of the concepts and technologies that make up the foundation of the Architecture Pillars, and it serves as an introduction to the contents provided through Chap. 4–10.

2.3 The Architecture Framework

2.3.1 Strata, Netlets, and the Design Repository

The Architecture Framework must provide ways to (i) guide the Network Architect to allocate the required network functionalities and (ii) assure the interoperability within families of network architectures.

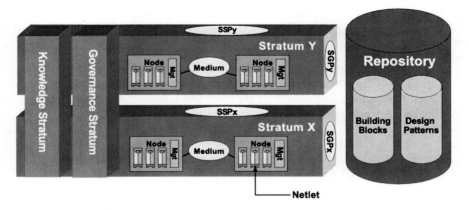

Fig. 2.2 High-level view of 4WARD Architecture Framework

As can be seen in Fig. 2.2, the following main components constitute this framework (see Chap. 4 for further detail):

- A Stratum is modelled as a set of logical Nodes which are connected through a Medium that provides the means for communication between the Nodes inside this stratum. This stratum encapsulates functions that are distributed over the nodes. These functions are provided to other strata through two well known interfaces (that can be also distributed over the nodes): The SSP (Service Stratum Point) that provides the services to the other strata located on top of the respective Stratum and to the vertical strata. Figure 2.2 shows Stratum Y using the services provided by Stratum X through SSP$_X$. The SGP (Service Gateway Point) offers peering relations to other strata of the same type.
- Strata can manage themselves. For example, when a routing service stratum is deployed, it organizes itself onto the physical infrastructure. The deployment will be in accordance with the specification of the logical nodes and the medium of the stratum, taking then into account the topology, capabilities, and resource status of the nodes and links in the physical infrastructure.
- Horizontally stacked strata (as shown in the middle of Fig. 2.2) are related to the transport and management of data across networks. Within such strata, Netlets can be considered as containers for networking services. They consist of functions/protocols inside a Node that are needed to provide the services. By virtue of containing protocols, Netlets can provide the Medium for different Strata, i.e. inside the same Netlet there could be functionalities that are related to different strata. Figure 2.2 shows such Netlets implementing media for different strata inside the same node.
- The two vertically oriented strata provide Governance and Knowledge for an entire network (i.e. a set of horizontal strata). The Knowledge Stratum provides and maintains a topology database as well as context and resource allocation status as reported by a horizontal stratum. The Governance Stratum uses this information, together with input provided via policies, to continuously determine an optimal

configuration of horizontal strata to meet the performance criteria for a network. The Governance Stratum also establishes and maintains relations and agreements with other networks.

The Repository contains the set of Building Blocks and Design Patterns for the composition of functionalities (i.e., to construct the strata and the netlets) for specific network architectures, including best practices and constraints to ensure interoperability between network architectures.

2.3.2 The Design Process

Evolution of today's networks including the Internet suffers from the inability to be extended in a consistent and reliable way while maintaining certain assured properties, such as security, quality of service, reliability even in the broader context. Much effort has to be spent for standardization, development and regression testing when introducing even minor feature improvements, before deploying them on a network-wide basis. Upgrading of a large installed base of network elements means a big technological challenge and financial risk to the network operator and service provider.

4WARD has succeeded in setting up a design process that in the future will enable new network designs to be developed, tested and deployed without impacting the installed network basis, when based on this 4WARD architecture framework and building upon the recent progress in network virtualization. The innovative 4WARD network design process leverages advantages of model-driven software engineering techniques and the experiences in design and composition of web services, based on OSGI principles [2].

As shown in Fig. 2.3, the following phases are considered in the design process:

1. **Requirements Analysis:** Starting from the business idea and requirements, the goal of this step is to decompose them into the high level functionalities that should be realized by the architecture to be designed. The output of this phase is mainly the identification of the macroscopic architectural view of Strata, a first draft of the main network components, and the specification of technical requirements for further refinement of the architecture.
2. **Abstract Service Design:** During this phase, the technical requirements and the high level functionalities derived from these will be turned into abstract functionalities and ways how they can be composed, following generic principles and design patterns. The result of this design phase is the specification of the Netlets operating at node level, and the Strata that constitute the distribution of functionalities across the network nodes.
3. The **Component Design Phase** focuses on the detailed specification and composition of the Functional Blocks (FBs) used to implement the specific functionality. This includes the specification of the interfaces, properties, and requirements/prerequisites of the FBs. The output of this phase is the detailed design of

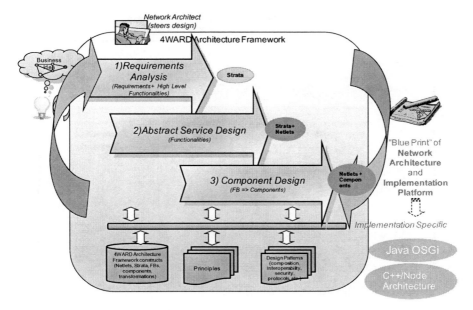

Fig. 2.3 High-level view of the 4WARD Design Process

the Netlets and software Components, which finally constitutes an "architectural blueprint" ready for instantiation on a network virtualization platform.

The entire design process is supported by an integrated design environment, which easily supports backtracking in iterative loops to redesign and improve the results of previous phases. In order to increase the reuse of architectural constructs and store the expertise and knowledge of the designing architect, an "architectural design repository" is used, which contains pre-built architectural constructs (abstract strata, netlets, components, functional blocks) as well as their derived instantiations, proven architectural design patterns on service and network composition, interoperability, security, etc.

2.4 In-Network Management

INM specifies two key architectural elements in order to realize distributed management within and across the network nodes: *Management Capabilities* (MC) and *Self Managing Entities* (SE). The MCs are encapsulations of management logic. The SEs are associated with a specific service and include relevant MCs for management of the service. Both elements are central to achieve autonomous behavior.

As part of the INM solution and design, algorithms have been developed for real-time monitoring, anomaly detection, situation awareness, and self-adaptation

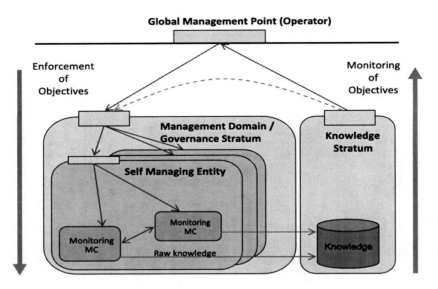

Fig. 2.4 INM relationship with Governance/Knowledge

schemes. The MC architectural element is the enabler of these algorithms. These algorithms provide best of breed mechanisms and patterns to address management tasks. They become important building blocks when designing networks. The 4WARD design process as described above includes an 'architectural design repository' which houses design patterns and network type building blocks available to the architect of the future networks. From a management perspective the algorithms developed for INM are key components of this repository which the architect can deploy as the need arises.

The 'management by objective' approach of INM is intrinsic to governance of networks and knowledge generation inside networks of the future. Both governance and knowledge are modelled as strata in the 4WARD architectural framework. Figure 2.4 shows management objectives being pushed downwards through the governance stratum, into the SEs and eventually into multiple MCs which carry out the tasks in hand. The MCs in the figure could for example implement a monitoring algorithm. The output of the monitoring algorithm is in essence unprocessed data. This is fed into the knowledge stratum and reasoned upon and more high level knowledge generated. This knowledge is then used, possibly fed back into governance if some modifications or tweaking are necessary or displayed at a higher level as feedback on the objectives which an operator applied to the network.

The algorithms developed and the management by objective approach which INM provides are key enablers in the realization of self managing, interoperable networks of the future.

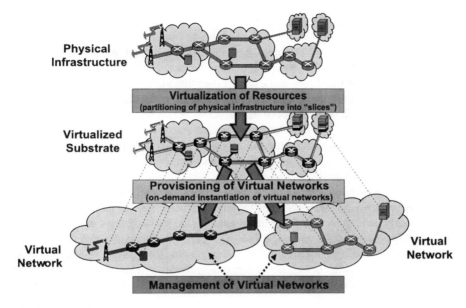

Fig. 2.5 Virtualization ecosystem

2.5 Network Virtualization

Virtualization has by now gained sufficient momentum as one of the key paradigms for future networking, as it has the potential to resolve the so-called "deployment stalemate" observed in today's Internet and foster the development of future networks paradigms. The straightforward use case for network virtualization is the scenario based on the decoupling of infrastructure ownership and virtual network operation.

The virtualization ecosystem encompasses three basic roles, namely (a) the infrastructure provider (having the capability to virtualize the physical infrastructure by partitioning them into 'slices'), (b) the virtual network provider (making the provisioning of complete end-to-end VNets by putting together 'slices' from the underlying infrastructure), and (c) the virtual network operator who is operating and managing a VNET. This is illustrated by Fig. 2.5. A service provider is then able to run specific services and applications on this VNet, which are then offered to end users.

Communication means between these actors and the definition of the respective interfaces constitute a cornerstone of the network virtualization architecture. This requires the specification of a formal virtual network description, allowing for flexibility, extensibility, scalability, interoperability and security. Since multiple business scenarios can be defined (ranging from vertical integration to a strict separation of roles), which imply different relationships of trust between them, the capability to define different levels of abstraction is also a key requirement. The 4WARD Resource Description Framework provides a language to describe virtual network resources and topologies, including all possible constraints that might be applicable in each case. An object-oriented data model was defined with four basic classes describing specific network elements, namely nodes, links, interfaces, and paths.

4WARD network virtualization architecture breaks with the traditional clear separation between a "dumb" core and a feature-rich edge in service provider networks. In this scenario, scalability will be a major challenge, particularly in terms of provisioning, management and control of virtual networks. A framework and algorithms for scalable mapping and embedding of virtual resources into the infrastructure, including discovery, matching, and binding were developed. Initial results suggest that the efficient construction of virtual networks from shared infrastructure at large scale is indeed feasible.

One of the most important features of the current Internet, global reachability and inter-networking, will surely remain a requirement in the future. This means that virtual networks, which by definition are separated and isolated from each other, will still need to communicate, although in a more controlled way. A concept for facilities to provide interworking between virtual networks, the Folding Points, has been developed, including the basic elements (Folding Nodes and Folding Links), as well as mechanisms for deployment using the virtual network provisioning framework.

2.6 Generic Paths

New mechanisms for data transport face contradictory requirements: large flexibility vs. uniform interfaces to all transport entities and efficient reuse of functionality are required. This can be partially achieved by new protocols only in end systems, but in general, an approach how to structure protocols both at the edge and in the core, at various "layers" is needed. For example, network management needs to identify, inside the network, data flows of different types; they should be able to give account of themselves (e.g., about their desired data rate) and obey a common set of commands.

To support such requirements, we focus on the data flow and its path as a core abstraction, along with a design process for a variety of path/flow behaviors. This process can incorporate new networking ideas; examples are network coding, spatial diversity cooperation, or multi-layer routing and is suitable for both end system and in-network implementation; the deployment is supported by the Architecture Framework.

The starting point for the 4WARD transport architecture was to find (1) a development model that can support reuse and flexibility, (2) a proper execution environment within a node (end system or router) with naming and addressing structure and a resolution scheme, and (3) the core functions and APIs necessary for a path, as generic as possible. Together, this is the core of the Generic Path architecture. It approaches issue (1) by using an object-oriented approach to define types of Generic Paths and to structure their interfaces; issue (2) by defining a set of constructs (namely, entity, endpoint, mediation point, compartment, hooks, and path) that describe the execution environment of instances of such path types; and issue (3) by selecting which operations should be possible on such paths (e.g., joining, splicing, or multiplexing). The concept shares some commonalities with OpenFlow, but concentrates on real-world necessities rather than on experimental usage; it also goes beyond merely modifying switching tables. To incorporate new networking ideas, all the relevant flows in a network share crucial commonalities and provide a common set of APIs with which to manipulate these flows. 4WARD's "Cooperation & Coding Framework" exploits such commonalities by addressing an entity that detects opportunities for turning on cooperation opportunities, like network coding, and can create the necessary path instances to setup a network coding butterfly. Mobility may be supported at different levels or compartments—and the realization of mobility at a session level is quite different from the realization of mobility at IP level, though they still share commonalities that can be defined through generalized mobility schemes. Thus, the GP framework allows the abstract description of a mobility process in terms of GP constructs, namely, entity, compartment, ports, path, and mediation point. Its realization can then resort to specific technologies adequate to the compartment we are considering in each case.

Based on this mindset, it becomes possible to develop powerful, custom-tailored path types. An example are path types for a Network of Information (described next), where the download of documents and the updating of location/caching tables can be tightly integrated and can access topology information to choose, for a document of interest, topologically close caches. Another example would be a path type to support the exchange of management information for In-Network Management entities, e.g., by compressing monitoring information more and more the further it is away from its source.

2.7 Network of Information

Today's networking is essentially about exchanging information between nodes. When accessing information, the request typically includes the host where the information shall be retrieved from, frequently in the form of a Uniform Resource Locator. This host-centric approach is often an obstacle for optimized transport of and easy access to information. Our approach to an information-centric architecture puts the information itself on the center stage. We take existing proposals that separate the host identity from the locator one step further by introducing information objects as first order elements in the network. In addition to classical scenarios such as content distribution, our work also encompasses scenarios that have so far not been discussed in the research community, e.g., the notion of real-world object tracking under the aegis of an information-centric architecture.

For the envisaged Network of Information (NetInf), we have developed an information model that constitutes a versatile and widely applicable framework for representing information in a wide sense. A clear split between the information itself and the location where it is stored is introduced. This eliminates the need for overloading locators and avoids putting them in the role of being an identifier and a

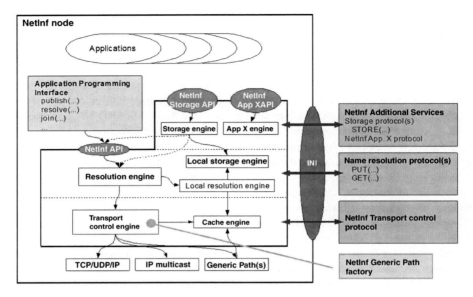

Fig. 2.6 NetInf high-level architecture

locator at the same time. The representation of the actual files containing the payload is called a *bit-level object* whereas the higher semantic level can be expressed by *information objects* that group or aggregate information.

The high-level architecture of a NetInf node is depicted in Fig. 2.6. The *NetInf Information Network Interface* (INI) at the right is the collection of NetInf protocols which are used to communicate to other NetInf nodes in the network. A uniform API exposed towards applications provides standard operations such as retrieving, publishing or updating information objects. This API can be extended with additional services. The Resolution Engine co-operates both with the local resolution engine when and if information objects can be found locally, but also with other remote resolution engines when such objects are stored elsewhere. Complementing the mobility schemes offered by the underlying transport, these mechanisms also provide a means to not only handle the mobility of nodes and networks, but also of information objects.

The NetInf Transport Control Engine is extremely flexible with regard to the transport mechanism that is utilized to transport the information objects or the requests. These transport mechanisms include, but do not mandate, the Generic Paths. Essentially, a set of adapted and optimized transport mechanisms applicable to information-centric networking are examples of specialized Generic Paths. The Transport Control Engine closely interacts with the Cache Engine which manages the caches that are used for short-term optimizations of data transport. The long-term memory of a NetInf system is provided by the (Local) Storage Engine. It uses the basic NetInf primitives to deliver and retrieve objects, while offering an advanced API that enables applications to manage the objects in the storage system, whether locally or remotely.

2.8 Conclusion, Reading Guidelines

In this chapter we have presented the 4WARD System Model, as well as the Architecture Pillars. The Architecture Pillars in turn point the key results of 4WARD, and a brief introduction was given to the concepts and technologies that make up the foundation of those pillars. The 4WARD System Model, through the definition of the Architecture Pillars, defines what can be understood as 'cornerstones' of what will be a more precise definition of an architecture for the Future Internet. Such an architecture will likely include also other building blocks in order to provide a complete and suited architecture for any type of network that would make up a part of the Future Internet.

The different elements and aspects of the 4WARD System Model are further described in Chaps. 4 through 10. Chapter 3 provides a description of the business, socio-economic, and regulatory aspects of future networks which gives important understanding of the interplay between business models, technical development, user needs, as well as regulation and governance. Chapter 11 provides a use case to apply the 4WARD System Model in order to analyze a specific business scenario as to derive a suitable network architecture for that scenario. Finally, Chap. 12 gives an overview of the various prototypes that have been implemented for the purpose of evaluating and demonstrating the 4WARD concepts and technologies.

References

1. M. Achemlal, P. Aranda, A.M. Biraghi, M.A. Callejo, J.M. Cabero, J. Carapinha, F. Cardoso, L.M. Correia, M. Dianati, I. El Khayat, M. Johnsson, Y. Lemieux, M.P. de Leon, J. Salo, G. Schultz, D. Sebastião, M. Soellner, Y. Zaki, L. Zhao, M. Zitterbart, 4WARD Deliverable D-2.1: Technical Requirements (Apr. 2009), http://www.4ward-project.eu
2. OSGi Alliance, http://www.osgi.org

Chapter 3
Socio-economic

Jukka Salo and Luis M. Correia

Abstract Non-technical drivers are addressed, when moving from the R&D stage to the real deployment of the technical and architectural innovations, grouped as: usage and services, socio-economic aspects, and regulation. Non-technical requirements are established, listed as twelve guidelines, which will have implications on network design rules. Four different scenarios are defined and developed, covering major aspects of both technical and non-technical areas: "Looking back from 2020: What made the old Internet break?", "Novel applications that are not possible with the current Internet", "Managing the Future Internet", and "Business models, value chains and new players". Six main drivers and challenges are presented in a scenario of evolution "Elephant and Gazelle", from the business environment viewpoint. Then, four different business use cases are addressed: *Network Virtualisation*, *New Ways of Information Delivery*, *Internet of Things* and *Community-Oriented Applications*.

3.1 Introduction/Setting the Scene

3.1.1 Overview

It is well understood that the development path of any industry or economic sector is significantly affected by the opportunities provided by the available technologies, the particular characteristics of its markets, and the directions and priorities of related government policies and regulations.

J. Salo
Nokia Siemens Networks, Espoo, Finland

L.M. Correia (✉)
IST/IT—Technical University of Lisbon, Lisbon, Portugal
e-mail: luis.correia@lx.it.pt

L.M. Correia et al. (eds.), *Architecture and Design for the Future Internet,*
Signals and Communication Technology,
DOI 10.1007/978-90-481-9346-2_3, © Springer Science+Business Media B.V. 2011

Fig. 3.1 Future Internet
ecosystem

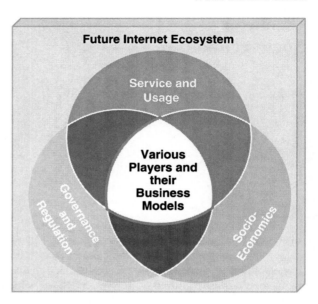

Previously, there has been a tendency to leave these issues to be handled separately, and the non-technical topics above have been addressed after the technology
had been developed [1]. In the case of the global networked society, this is not a
desirable approach. Moreover, the different problems related to those topics are at
such a high level that new debate and interdisciplinary research between technology and policy experts is urgently required. The take-off and success of the Future
Internet will be closely linked with actions taken in all areas of the Future Internet
ecosystem (Fig. 3.1).

In order to properly cover the business aspects of the Future Internet, the impact
of the **non-technical drivers** has to be studied [2] (grouped in three areas), when
moving from the R&D stage to the real deployment of the technical and architectural
innovations.

The non-technical drivers are grouped as follows:

- **Usage and Services:** any new technology, even if it is excellent, can only have a
  market success if it satisfies current or potential user needs in a sustainable way.
- **Socio-economic aspects:** social trends and issues can significantly boost or hamper the success of an innovation.
- **Regulation:** the traditional telecom field is strictly regulated, as opposed to
  Internet-type data networks, different stakeholders having different views on the
  regulation of future networks.

The new generation of Internet technologies will create market opportunities for
small- and medium-sized businesses, and allow them to effectively bring innovative
services to subscribers. This will enlarge the overall market, and open up many new
opportunities for different players.

3.1.2 Usage and Services

The theme Usage and Services links technology with the customer and his/her needs and requirements. The target of this theme is to understand how technologies might be used by customers, and what their impact on the existing service models might be.

Today, services can be divided roughly into two layers: application and connection. Application services are those that can be used directly by customers, like telephony or web browsing. Residential customers are normally not aware of connection services as a key enabler to application ones and, hence, do not care about the technology they use to be connected to a network. It is not necessary for them to understand DSL, UMTS or even "simple" GSM or PSTN; in addition, they are not interested in solutions to technical problems, like coding or encryption. Business customers have also little interest in connection services, because more and more companies are outsourcing their infrastructure to reduce the complexity in, e.g., managing network connectivity. Besides the two service layers, there are also network functions providing generic functionalities, like AAA, VPN, or firewall.

Both application and connection service providers need a good understanding on the users' behaviour for dimensioning and planning the network and related parts. The modelling of the usage and services is influenced by a number of technical parameters (e.g., number of users, bandwidth, and required CPU load), social context (at home, work, etc.), mobility association (on the move, in a moving environment, etc.) or location type (building, park, etc.).

A more detailed look into Usage and Services activities will also have to consider the customer's viewpoint on design aspects. In the past, various solutions were too difficult to be useable for customers, or did satisfy customer needs only to a minor extent.

3.1.3 Socio-economics

Market success is never solely driven by technology. Technology is needed as an enabler, but it takes time for a new technology to be exploited, to become easily usable, reliable, affordable and successful. The influence of a new technology on existing and emerging business scenarios needs to be investigated, as this may also reveal new requirements for the new technology.

A great deal of Socio-economic items is influencing today's use of web 1.0 and 2.0 services, and people's way of living, while all the events that modify society and life habits usually impact also local or national economies. These events can either be worldwide macro phenomena (e.g., international trade treaties, changes in the physical environment, ecological changes, or a worldwide financial crisis) or small and local changes (like closing a factory or moving a local manufacturing plant to the Far East). The Future Internet will play a dominant role in the society, as new and far closer relationships are expected to be set up among businesses, public sector, and consumers, based on reliable communication networks and services.

In the following, a number of the key Socio-economic factors, which will have impact on the development of the Future Internet, are discussed.

A key factor is **affordability**. People must feel that the expenses incurred for a new enabling technology are really "useful" expenses. This factor may strongly fluctuate over time; for instance, a Forrester Research report from November 2008 [3] says that 58% of adult Americans are spending less money now than they did a year earlier as a result of the economic situation.

The second key factor is the world's **population structure**, which will strongly affect terminals' and services' development, managing and provisioning. In developed countries, 20% of today's population is aged 60 years or more, and by 2050 that proportion is projected to be about 1/3 [4]. Provided that the advantages of the new services are clearly visible, and usage is really intuitive for elderly people, new technologies enabling a better lifestyle or compensating restrictions coming with old age should become very successful.

The third key factor is the opportunity for individuals belonging to lower social classes to change their social and economic status by **improving education, qualification and employment** opportunities thanks to the growing importance of communication networks. While in developed countries the budget for communication in private households is expected to remain a stable fraction of the total budget, real growth will happen in developing countries, where the available budget is going to rise when basic needs like alimentation and accommodation are fulfilled.

Climate change and energy use are issues likely to have a strong impact on the world economy and way of living. A significant amount of CO_2 emissions result from the delivery of goods and transports of people for business and leisure purposes. The energy and climate problem will also impact on economic growth, as increased expenditures will be required to mitigate the effects of climate change. The Future Internet is required to produce innovations to reform the use of energy and to change social and economic habits.

Another key factor is the growing demand for a **seamless mobility** and for being "always connected". These needs may range from just talking or sending and receiving simple messages up to having access to music, news and entertainment services to feel in touch with the world for different purposes.

The **Internet of Things** will also spread, as many devices will become software controlled and able to interact with other devices, applications and persons: appliances in the kitchen; gadgets in the living room; lighting, heating, cooling, watering, and draining facilities in the building; the array of thermometers, scales, and health meters; and recreational, educational and entertainment equipments.

There is also a **dark side of all these key factors**: a lot of questions about **privacy, security, access control, identity stealing** will arise and create manifold concerns to private persons and businesses.

3.1.4 Regulation

Today, the Internet has already a huge social and economic impact in all countries, and this will even increase in the future. The Future Internet will be the basis for

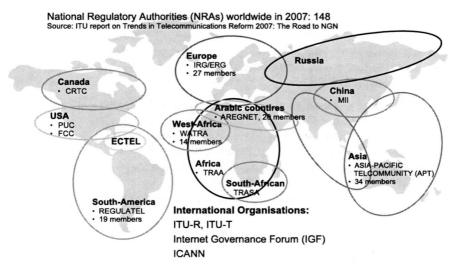

National Regulatory Authorities (NRAs) worldwide in 2007: 148
Source: ITU report on Trends in Telecommunications Reform 2007: The Road to NGN

Fig. 3.2 The international regulator landscape (source [7])

more efficient public services, and will improve the relationship between citizens and their governments. Solutions, which empower citizens to participate in healthcare processes remotely and facilitate remote monitoring and care, or which will allow social interaction without travel, are only examples of what can be expected. The Future Internet will also play an important role in exercising democracy, and perhaps it is not that far in the future that voting in general elections can be done using Internet. On the downside, more and more communication and information technology will be used for some kind of surveillance, and many expect that this will increase privacy erosion. It is clear that the Future Internet will become so important that governments want to know how it will be run. Rules, policies and laws become important, and interaction is needed between different communities to make these rules reasonable [1, 5].

Since the Internet infrastructure and Internet markets are global by nature, global approaches are needed also in the area of policy, governance and regulation. However, the world from this perspective looks quite fragmented as shown in Fig. 3.2: in 2007, there were 148 National Regulatory Authorities worldwide, which are responsible for the economic regulation of the communications markets, and for the supervision of the technical operability and safety of the communications networks in their countries. In addition, there are a number of regional organisations, which are trying to harmonise the rules in a certain regional area. In the European Union, for instance, the European Regulators Group (ERG), which was created in 2002 [6] (the Body of the European Regulators (BEREC), established in 2009 and started in January 2010, continues the work of ERG) is to provide a suitable mechanism for encouraging cooperation and coordination among national regulatory authorities and the European Commission. In other geographical regions there are equivalent organisations.

Regarding the potential issues and challenges in the regulation of future networks, central or peripheral to those networks, a lot of attention has to be paid to the fact that the number of these issues is growing at a rate that both legislative bodies and executive forces may have difficulties to follow up. Among these issues and challenges are [1, 5, 8, 9]:

- **Variability of national rules:** possible temptations towards censorship, governance, lawful interception, government inspection, standards, etc.
- **Regional and global markets:** telecommunications markets are increasingly becoming regional and global, global thinking being needed from all parties.
- **Large number of actors:** decisions tend to take too long and results tend to be the least common denominator without visionary aspects.
- **Different views on the future network:** is focus needed on the infrastructure or on the service component?
- **Different areas of convergence are regulated differently:** telecoms, broadcasting, broadband access, services are at stake, hence, how to harmonise the different rules for the Future Internet?
- **Privacy and security:** public interest seems to conflict with commercial interest.
- **Controlling responsibility:** responsibilities for the legality of information sharing and the legal usage of network resources have to be clarified.
- **Capability to trace back messages to their source:** IP traceback facility is needed to ensure the non-repudiation of the message originator.

3.2 Non-technical Requirements

A continuous dialogue between the technical and non-technical work areas is needed when developing the Future Internet, ensuring a meaningful evaluation of the related concepts and techniques. Major guidelines are presented below [10].

Service usage is expected to evolve from the traditional notion of using a service for short periods of time, and for specific purposes, towards the usage of an infrastructure that intelligently supports the daily life of users in a transparent manner. Also, as services and networks themselves will continue to evolve, future network technologies and protocols need to flexibly adapt to upcoming new requirements. This aspect is also an essential guideline in the European FP7 project TRILOGY [11], where this guideline is named "Design for Tussle".

The requirements analysis started from the traditional service model, but new emerging applications with new usage patterns (e.g., users producing content) will certainly have implications on the network design rules. New business models will emerge and change the value chains, impacting the network and service operator business, while, at the same time, creating possibilities for new players.

The following guidelines for the development of the Future Internet technologies were identified to be relevant from the Usage and Services' viewpoint:

- **Guideline #1:** Future Internet technologies shall support a broad range of innovative services, delivered to human customers as well as to machines or virtual objects: as the details of innovative services cannot be planned in advance, technologies shall flexibly adapt to emerging requirements.

- **Guideline #2:** Future Internet technologies shall support the existing business models, as well as the emerging, ones; they shall enable new players to introduce commercial and non-commercial services, without disrupting existing ones and without jeopardising their evolution (i.e., guaranteeing some kind of backward compatibility).
- **Guideline #3:** Future Internet technologies shall support service provider requirements for managing their operations, including information about users and usage patterns, in compliance with legal rules respecting privacy and competition.
- **Guideline #4:** Future Internet technologies shall support mass market customer requirements, implying the provision of a satisfactory quality of experience for the average customer and a high quality of experience on demand.

Internet-based services will become ubiquitous, and will be underlying to all social and economic infrastructures, thus, the society at large will depend on the Future Internet. Ultimate requirements will apply to safety, reliability, and dependability of this critical infrastructure. Attention shall also be paid to potential network-based criminal activities (e.g., fraud, spam, and sensitive data stealing); in the end, the achieved protection level has to be traded-off against its cost, in terms of both financial aspects and potential bureaucratic innovation barriers.

The success of Future Internet technologies will depend on how well they fit to the overall Socio-economic context. World population will be close to 9 billion by 2050, elderly being a 1/4 of the population, while children aged up to 15 years old will be only a 1/5. For this reason, the demand for traditional, reliable and easy-to-use products and services will grow. Services that allow entertainment via web instead of *going to* a place, nursing via web instead of *having a person* at home, remote control instead of *having medical check* at home, etc., will be required. Also, to support the 'Networked Everyday Life' in a confident way, a high level of reliability will be important.

The Future Internet will also play a critical role in fighting against climate change, by reducing energy consumption due to travelling, for example. But attention must also be paid to the energy consumption by Internet infrastructures, as it is by far not negligible and needs to be justified by significant savings in other areas.

The following guidelines for the development of the Future Internet technologies were identified to be relevant from the Socio-economics' viewpoint:

- **Guideline #5:** Future Internet technologies have to support safety-critical applications. Network and service availability, with a satisfactory performance, needs to be secured under all circumstances for lifeline services; with respect to other services, network dependability shall allow their usage in the critical processes of daily life. Also, personal privacy, as well as the protection against network-based fraud, spam, and other criminal activities, must be assured.
- **Guideline #6:** Future Internet services shall address huge societal challenges at the verge of unsustainable population density in some regions, and also of ageing societies.

- **Guideline #7:** The Future Internet shall relieve the strain on the environment and shall be a "Green Technology".
- **Guideline #8:** The Future Internet will have to allow communication not only among persons, but also between persons and things, and among things, supporting the "Networked Everyday Life" and impacting on many aspects of the social life.

The different problems related to Regulation are at such a high level that a new debate and interdisciplinary research between technology and policy experts is urgently required. It is clear that whatever structure of the future telecom regulation is adopted, all countries will need to pay much greater attention to the need for increased coordination of policy directions and regulatory activities, both across industries and sectors, and with other countries.

Within Europe, the representatives of the European Commission have stated it clearly that any further redesign of the architecture of the global networks will have to respect the basic characteristics of the openness, interoperability and end-to-end principles. The adherence to such basic principles is clearly an area for international cooperation at both technological (saying what is possible) and policy (saying the requirements) levels.

The regulatory issues of Internet cover the areas of infrastructure, security, stability, privacy, intellectual property rights, national sovereignty (country domain names for example), etc. [12]. There is a strong ongoing debate about the governance of Internet, and several parties have questioned the current role of ICANN. There is no good understanding how to handle different conflicting interests, as well as legal and cultural limitations. When developing the Future Internet, it is important to identify the items to be governed.

The following guidelines for the development of the Future Internet technologies were identified to be relevant from the Regulation's viewpoint:

- **Guideline #9:** The technical development of the Future Internet shall monitor ongoing discussions about policy, governance and regulation. A feedback loop towards technology development and business modelling shall be maintained. The regulatory matters relevant to Future Internet technologies and services include issues that are directly related to the network (e.g., topology, protocols, addressing, and QoS), as well as others regarding naming services, exchange of information, global coverage of service provisioning, and privacy and security, have to be paid attention to.
- **Guideline #10:** Technical work on the Future Internet has to follow the decentralised and collaborative process of the underlying technological development and core resource management ("Internet community style"). This work shall result in a distributed/decentralised open architecture, which principle has been proven effective to achieve an interoperable, functional, stable, secure, efficient as well as scalable network. By this, any type of network anywhere can be included and be made publicly available.
- **Guideline #11:** Technical work on the Future Internet shall respect the open, nonproprietary nature of the core Internet standards. Protocol specifications shall be

available to anyone, at no cost, thus, considerably reducing barriers to entry and enabling interoperability.

- **Guideline #12:** Technical work on the Future Internet shall support competition and innovation. These market mechanisms, supported by the liberalisation of markets, have by and large enabled the current development of the Internet.

3.3 Scenarios for Evaluation of the Major Driving Forces

3.3.1 Definition of Scenarios

Four different scenarios were defined and developed taking the key drivers and challenges of the Future Internet into account. This set of scenarios was defined in order to cover major aspects of both technical and non-technical areas [2].

Scenario 1 ('**Looking back from 2020: What made the old Internet break?**') outlines which technical and non-technical developments will be decisive for the understanding that the smooth evolution of the existing Internet concepts will no longer be applicable in the communication world. This includes the analysis of infrastructure problems, innovation restrictions, and the limitations in economic incentives. The outcome of this scenario depicts a set of problems that may or will make the current Internet break: usability problems, ignorance of security issues, and human communication problems; network accessibility problems for dependable communication; misuse of identity information; implementation and product deficiencies; growing costs for fixing obvious problems to maintain a minimum network reliability making network operation economically infeasible.

Scenario 2 ('**Novel applications that are not possible with the current Internet**') identifies the challenges that will be posed from conceivable new applications to the Internet, and that are not possible, or are very difficult, to implement using the existing Internet concepts (from economic or technical viewpoints). The developed ideas include: context awareness; micro-service provider; personal networks; augmented reality; resources on demand; 3D video.

Scenario 3 ('**Managing the Future Internet**') concentrates on network management issues that come up with broadening the traditional one-stop-shop operator to an environment with several competing and partly collaborating network operators and a multitude of service providers. The objective is to have self-managed networks, which are built-in at design time. Major themes covered within the scenario work are: the blurring boundaries between operators and other players in a future Internet; the growing complexity of infrastructure and services, and the associated need to find new ways of network/service management; the new capabilities provided to operators, based on innovative future Internet technologies.

Scenario 4 ('**Business models, value chains and new players**') focuses on the non-technical aspects of the Future Internet. It evaluates the impact of social, economic and political trends on the telecom business, in order to work out the most

Fig. 3.3 Extreme scenarios characterised by six drivers with uncertain development

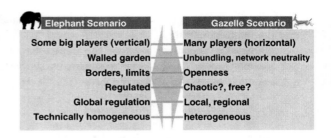

decisive elements that will govern the future business environment. The scenario description includes the study of a couple of issues in the context of an Internet break: change of the market balance; opportunities for new business players; regulation requirements to enable fair and reliable market conditions; cooperation strategy for established actors.

3.3.2 Business Environments

A major result of the scenario work is the compilation of the main drivers and challenges of the Future Internet business. For most of the drivers, there was clear agreement about their future impact and relevance. Only six of them give reasonable different estimations on their future impact and relevance. These basic questions are:

- Will the Internet arena be dominated by a limited number of big players, or, on the contrary, is it more feasible that a multitude of specialised small companies will satisfy the increasing demand for individual services?
- Will centralisation (e.g., big server farms) or decentralisation (peer-to-peer networks) determine the direction of future developments?
- What will be the main obstacle for growth: regulatory intervention, compatibility problems of technical solutions or a mismatch in market power?
- How can the global usage and accessibility of the Internet (the origin of its success) be assured under different market environments without global regulation?
- Will heterogeneity in technology accelerate or delay technical innovation? Is the coexistence of multiple heterogeneous platforms (which may be operating on the same physical system, but separated by virtualisation) a good alternative to one standardised platform with a lot of complications due to increasing complexity?

A picture of two scenarios describing extreme market positions from a global view is given in Fig. 3.3. The 'Elephant' scenario is characterised by strong forces that preserve a given market regime, and allow changes of market forces only in slow-motion. In contrast, the 'Gazelle' scenario predicts the coexistence of many players in a highly dynamic market.

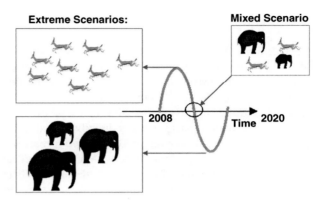

Fig. 3.4 Evolvement of markets in the regulated environment

In the 'Elephant' scenario, the market is distributed among a few big players, who are protecting their market position like a 'walled garden'. The 'walled garden' approach allows market players to better control their assets, and provides an increased security and quality performance for the subscriber by tight access control and special agreements between the partners in a 'walled garden' system. Clear borders will limit the open options of third parties. If only a few big players control the whole market, they will likely use standard technologies within their networks. By this, homogeneous technical solutions become accepted.

In all of the considered items, the 'Gazelle' scenario is the opposite of the 'Elephant' one. A big number of small players will struggle for best solutions. Frequently new players will appear and old ones will disappear. Open platforms ensure low entrance barriers for new players. In this context two questions arise: "Who will provide the open platforms?" and "What are the rules of the game for platform operators and are they out of the scope of the 'Gazelle' scenario?". The presence of many, about equally sized, players will guarantee fair market conditions in the 'Gazelle' scenario. But there will be a risk whether such a free and probably chaotic market can satisfy the basic requirements of customers (coverage, emergency calls, etc.).

The regulator will be the key player in both scenarios. On the one hand, the mandatory requirements have to be assured by regulation, like lawful interception and emergency call realisation; the regulator has to take care of discrimination-free access to network services for all market players, and to control the fulfilment of prescriptive geographical coverage requirements for certain services. On the other, the regulator is responsible for a balanced and healthy competition. The regulation is asked to coordinate the market in case of unequal concentration of power. Regarding the 'Elephant' scenario, regulation can start to change the picture and prepare a framework supporting the development towards the 'Gazelle' one. If an open and free environment with less regulation is achieved, market concentration might start again as a result of economics of scale. Figure 3.4 depicts how such a cyclic process could evolve over time in a regulated environment. The style and speed of technological innovation may depend on the phase of the market cycle.

3.4 Business Use Cases

Four different Business Use Cases were developed: *Network Virtualisation, New Ways of Information Delivery, Internet of Things* and *Community-Oriented Applications*. The first two scenarios correspond to the usage of technologies developed within the project, hence, linking the development of new technical solutions with their use and the business models that will support them; given the nature of these technologies, their impact is not felt directly by end-users, but rather targeting companies and other business entities. The last two scenarios are targeting end-users, and the way that technologies developed within the project can directly impact on them; this enables the bridging of the project's developments to end-users. In the following sections, the Business Use Cases for *Network Virtualisation* and *New Ways of Information Delivery* are described and analysed, then short descriptions are given for the other two.

3.4.1 Network Virtualisation

3.4.1.1 Overview

Virtualisation is a broad concept in the information and telecommunication area, dealing with the hiding/sharing of physical, as well as virtual, resources. There are already several different virtualisation concepts adopted in practice, which target operating systems, hardware, CPUs and embedded systems, networks or storage. The general advantages of sharing resources between different applications are, for instance: reduced number of equipment devices, commoditisation of resources (less proprietary systems), reduced complexity in the management of resources, reduced time needed for deployments using the virtualised infrastructure and flexibility in usage.

The economic aspect lies in the optimisation of needed resources, and therefore in the reduced Total Cost of Ownership. Besides, there are several aspects that have to be taken into account when considering network virtualisation, e.g.: better planning of the needed shared resources; additional management for resource sharing; integration of specialised resources requiring higher efforts; and operation and maintenance requiring additional debugging mechanisms.

There are also additional advantages for the various kinds of virtualisation, such as license sharing or power reduction due to the server virtualisation.

Figure 3.5 gives an overview of the network virtualisation ecosystem. One can observe that a multitude of different potential interrelations between players allows for the setup of a lot of potential service provisioning scenarios. There are essentially five main players involved in the virtualisation business ecosystem [13, 14]:

1. The **virtual network user** (not in the figure) is the end customer accessing applications over virtual networks.
2. The **Service Provider** (SP) deploys services or applications on virtual networks.

Fig. 3.5 Network
virtualisation ecosystem

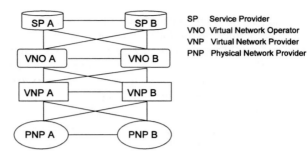

3. A **Virtual Network Operator** (VNO) operates, maintains, controls and manages the virtual network, which can be done once they have been composed or built from virtual resources by the virtual network provider.
4. A **Virtual Network Provider** (VNP) requests virtual resources from various infrastructure providers, composes virtual networks and offers them to virtual network operators.
5. A **Physical Network Provider** (PNP) extracts a part or the entire virtual network from its own physical resources and binds them on behalf of the virtual network providers for later use by the end users and service providers.

3.4.1.2 Business Perspectives

The area of Network Virtualisation as a service offering is a new business area in the Internet, which provides a possibility to optimise the operational area of a business player. However, the current service offerings of network operators will be affected and revenues will decline, as the service provisioning will be reduced to provisioning the virtual network connectivity.

It can be expected that all areas of a network operator's business process will be impacted [15], therefore, their resources need to be revised. These resources include personnel for planning and operating systems, software for planning and operating of virtualised infrastructure (Operations Support Systems), business process oriented support systems (Business Support Systems), or hardware for virtualised elements (e.g., increased CPU power or memory).

The following markets or market opportunities might be impacted or created:

- Supplier market for telecommunications hardware and related software, e.g., operating systems.
- Software developer market for developing networking protocols and management solutions.
- Network operator market for providing connectivity to customers and operators.
- Service provider market for highly optimised service delivery methods.
- Business customer market for LAN-like services across multiple physical and decentralised networks.
- Residential customer market with evolving social requirements.
- Training market for enabling employees to manage the new virtualised networks.

The general economical aspects of virtualisation and the impact on a player as such are outlined in the following. For a certain business player, the impact could vary as it could take several roles/responsibilities in the value chain (e.g., an operational split between network and service of a player might increase the internal costs between the two now independent business parts). Costs could become lower as more and more parties utilise unused resources in the network and costs are shared among them. This operational optimisation will make sense only for operators taking more than one role in the virtualisation value chain.

A possible optimisation area is the protection/isolation of physical resources through a virtualised access. This may reduce contractual penalties caused by bad configurations, because service networks are isolated and protocol interworking is reduced. This can be relevant for a business player, but not for a specific virtualisation business role. Each VNP might offer different protocols or allow different protocols to work in his environment. Also, the PNP might offer virtualised resources with specific characteristics, e.g., the support of MPLS (Multiprotocol Label Switching).

The transition from one network layer technology to another is an additional optimisation area. Virtualised networks might provide the possibility to run a new network technology in parallel with the old one, without changing the hardware in the first place. Virtualised platforms might provide only computing power to VNPs or VNOs, and they could run any protocol they want. It is also imaginable that the PNP limits the choice of protocols, and offers only limited possibilities for VNPs or VNOs.

An analysis of the value chain is also of interest.

Michael Porter developed the "Five Forces of Competition" qualitative methodology [16]. It is meant for assessing the challenges to the attractiveness of an industry and the impact of changes on a company's own position. This systematic approach was used to transparently evaluate a sample of business cases for the Future Internet. Starting from each business role, there are four different markets in the virtualisation value chain. In Fig. 3.6, Porter's Five Forces are applied on each of the business roles. It can be seen that the value chain is coupled by the bargaining of suppliers and buyers, because both are the converse to each other: for one role a player is the supplier, while for the other role the same player the buyer.

Rivalry will exist in the different aspects, and network virtualisation has to be seen as a new entrant in the sense of Porter's Five Forces. Therefore, today's existing industry is covered in the force "Threat of substitute products or services". In the following, the conclusions for each role are outlined.

- PNP—The number of new entrants will be relatively limited due to high investments and fixed costs. Regarding the bargaining power suppliers, standardisation will play a key role. On the other hand, standardisation limits opportunities for product differentiation, but this might have less impact for products like processing power and bandwidth. In addition, the competition with the existing telecommunication (technology) provider will be high. Combining the previous facts and pointing again on the product differentiation difficulties, rivalry among the players will be high.

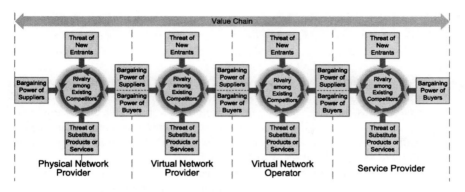

Fig. 3.6 Rivalry in the value chain of virtualisation

- VNP—The number of new entrants will be high, because of the relatively low investments and fixed costs in this business area. Standardisation will play a key role, but is not as important, because product differentiation will depend also on the coverage a VNP can provide; in addition, coverage will have influence on the profitability. Similar to PNP, the threat on products and services from the existing telecommunication (technology) providers will be tough. Overall, the rivalry among VNPs will be high and depend on the player and its ability to provide unique and interesting interconnections.
- VNO—It could be assumed that the number of players will increase as the complexity to manage a network decreases. On the one hand, only a limited number of players take the majority of the market share, and the rest consolidates; on the other, the Internet market shows that consolidation could take a long time. This has a huge impact on the cost side, because economies of scale will be difficult to reach in a highly diversified market. Potential new entrants could be software developers of the management software. The virtualised products of the VNO must be defined in a way that their implementation and also availability in end-user devices is ensured. A very successful example for the broad acceptance and availability are the Internet protocols like IP, DNS and HTTP. The previous factors lead to the conclusions that the rivalry among players will be high and the bargaining power of customers will increase.
- SP—The number of SPs could vary from a relatively restricted number to tens of thousands (in case of network providers) or above. Correspondingly, the number of customers will be reversed. In the beginning, only network operators and application service providers are important, therefore the number of players will be relative low. The entrance of new players can be assumed. Comparable to fixed or mobile VNOs today, creditable brands will go in this direction like a supermarket running its own mobile brand. The key criteria will be the willingness to run a business and the relationship with potential customers. Rivalry among players will be high, if the target for a simplified network organisation could be reached.

New services will appear in the area of virtualisation of resources for network and service providers, but it can be assumed that they will not directly face the end-customer. Instead, new application services exploiting virtualisation features may be offered by network and service providers, and this may include the exploitation of virtualised processing power, bandwidth, and protocol specifics like addresses or labels.

The optimised utilisation of resources will be a result from the new network services offerings. Based on the virtualisation and the appreciated increased usage of existing resources, the usage itself could be promoted. Support of a bigger number of players, especially by enabling micro providers, will enrich potential usage scenarios. Finally, improved service offerings will change the behaviour of end-customers, and will increase the utilisation of services.

Virtualisation will impact on many socio-economic issues, e.g.: reducing capital expenditures with better resource utilisation; increasing the performance of networks when using 'always best routing'; reducing the maintenance down times; and increasing security due to that virtual networks give less chance for attacks.

Policy, Governance & Regulation will also be impacted by virtualisation. A key regulatory issue, namely Interconnection, is discussed in the next section.

3.4.1.3 Regulatory Perspectives of Interconnection Issues

According to ITU surveys [8], interconnection related issues are ranked in many countries as the most important challenges in the development of a competitive marketplace for telecommunications services. The purpose of an interconnection regime is to benefit users by encouraging competition that will lower the price and improve the scope and quality of services. For competition to be successful at maximising consumer benefits and innovation in telecommunications market, network providers must have the opportunity to access all customers, even those customers that are connected to the networks of their competitors.

In the Internet of today, interconnection usually is implemented on the base of voluntary agreements between IP Service Providers. These freely negotiated arrangements have resulted in a richly interconnected Internet, and do not depend on regulatory obligations. The Future Internet, which will be more complex having several different technical and administrative domains (e.g., virtual networks) and carrying new types of traffic, raise concerns on whether the old approaches are still sufficient in the future.

For identifying potential interconnection issues in the Network Virtualisation concept and ecosystem, several interconnection cases were created and analysed [10]. They are based on the different interconnection needs of a Virtual Network (VNet), as illustrated on the left side of Fig. 3.7. A VNet could be interconnected with another one, an interconnecting VNet (ICVNet), a Physical Network (PN) and a Service Provider System (SPS). In the use case illustrated on the right side of Fig. 3.7, the two VNets have been provided by the same VNP, but VNet A uses the resources of PN 1 and VNet B uses the resources of both PN 1 and PN 2. The VNOs have to agree on the capacity of the interconnection link, the different classes of QoS used on the link, who pays to whom, etc.

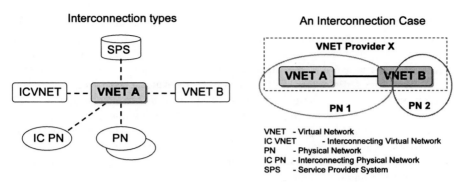

Fig. 3.7   Different interconnection types and an interconnection case (example)

The analysis resulted in a list of potential interconnection issues in virtual networks. They were assessed against the criteria of economic efficiency, which is one of the main contributors to the consumer welfare. Economic efficiency is defined as the best use of resources (allocation efficiency), least cost production (productive efficiency), and incentives for innovation and investment (dynamic efficiency). These dimensions of efficiency may conflict so that determining the optimal charging model for interconnection may require balancing differing impacts [17].

The analysis of the interconnection use cases resulted in a list of potential issues with respect to the Network Virtualisation, which are described in what follows:

- Location of Points of Interconnection—The cost for routing a message depends on the number of routers that a message has to go through. Using the 'hot potato' routing, costs are minimised by handing messages over to another network as close as possible to a network's own retail customer. Alternatively, an interconnecting operator might handover a message to another network at the closest point of interconnection in order to save the capacity of its own network. From the perspective of the total cost for transporting a message, this practice may not be the most optimal one.
- Interconnection Capacity—An incumbent operator could prevent competition with new entrants by providing insufficient amount of capacity for the interconnecting networks, or by providing only low quality interconnection capacity. Playing with the interconnection capacity could also be used to discriminate between different service providers. For a new entrant, interconnection with incumbents or with other new entrants can be seen as an improvement of its competitive position (enable the business), or even as a prerequisite to start any business. For interconnecting virtual networks, capacity can be made available in a physical network or in an interconnecting virtual network. Service Level Agreements are needed for interconnections, on either the virtual network or the physical one level. Investing in capacity has to be profitable.
- Quality of Interconnection—Future networks will support different classes of service (QoS). This will be true also with respect to interconnection, and a lot of harm would be caused to end-users if interconnection would be of low quality.

Providing interconnection only on a low quality level would benefit mainly an incumbent operator, who has the major share of the customer base. The high quality of interconnection implies also the availability of sufficient interconnection capacity to avoid congestion. Because different classes of service may have different prices, the related traffic needs to be measured class by class. Offering high quality interconnection services could be a means to compete by a VNO; other VNOs could be willing to pay extra for high quality interconnection.

- Interconnection Discrimination and Network Neutrality—The discussion on the Network Neutrality relates with the following three contentious issues: discrimination, blocking of user access to content, and access. Potential areas for discrimination regarding interconnection are:

  (a) Discrimination by the infrastructure provider between different virtual networks and/or interconnecting virtual networks.
  (b) Discrimination by the VNO, by the interconnecting virtual network operator, between different VNOs, and between PNPs.
  (c) Discrimination by the VNO between third-party content providers and its own subsidiary content provider (a form of first-line discrimination).
  (d) Discrimination by the VNO between various content providers (a form of second-line discrimination).

Insufficient network capacity for interconnections will be fatal to the prospects of competition, since the resulting network congestion can be a deadly anticompetitive barrier. Blocking of user access to content, applications and services would erect a barrier in either one or both links between an end user and the content provider of his choice.

- Addressing, Numbering and Number Resolution—Interconnection of networks (including virtual networks) implies that their addressing is based on the same addressing scheme, or that there are mechanisms for the address translation at the point of interconnection. Since it is quite unrealistic that the addressing schemes in different types of networks (cellular or other) would be the same, joint approaches and rules are needed for the number translation and for allocating address or number space for the different players.
- Cost of Interconnection—Cost of interconnection depends on the capacity of interconnecting links, the required QoS level, and related operating costs. The current interconnection at the IP level does not distinguish between the different classes of service. In the future, however, the network resources consumed will vary with the required QoS, and these costs must be taken into account in the interconnection arrangements; the cost burden has to be appropriately shared between operators and their respective customers.
- Variability of Interconnection Types—Interconnection of the virtual networks implies interconnection of the related physical networks, irrespective of their type and technology. Since the different interconnection models (including charging) apply for different types of networks, resolving interconnection issues may become challenging. For example, interconnecting a cellular network and an Internet network for the VoIP service may be needed. There are several types of

interconnection also on the virtual network level, including peer-to-peer interconnection between two virtual networks and between a virtual network and an interconnecting virtual network. End users need to have equal ease of access to the competing networks, irrespective of the interconnection type.

Potential regulatory actions in the context of the Network Virtualisation concept and Business Use Case are listed below (an action may be relevant to several issues listed above):

- An efficient interconnection model shall be imposed for ensuring the availability of the sufficient amount of interconnection capacity and the timely provisioning of interconnection services. Also, it is important that interconnection is available to the services irrespective of who is providing them.
- Fair charging models have to be imposed, and they can be based on the offered interconnection quality. The cost based charging is often recommended for enabling access to the markets by the new entrants. Costs need to be identified and made transparent, in order to prevent a dominant operator from demanding non-reasonable high prices for terminating calls that originate from a new competitor's network. The transparency of pricing of the interconnection capacity could be achieved, e.g., through the structural separation. Also, the existence of several infrastructure providers would keep the prices reasonable, and shall be promoted.
- Open access policies to the virtual resources shall be followed, and can be promoted through the functional separation, wholesaling and unbundling the interconnection resources. Unbundling of the essential facilities is an approach to boost competition in various telecommunications service markets.
- Global addressing and numbering schemes shall be promoted across country borders. However, since it is unrealistic that the schemes would be exploited everywhere at the same time, the address and number translation function has to be supported at the points of interconnection. Also, the address and number portability has to be supported at all levels of the network.
- Harmonisation of the regulatory policies for the different types of networks shall be promoted. Standards have to be specified for ensuring the interoperability between the fixed, mobile, internet and broadcasting networks. Similar regulatory rules shall be applied on each of these sectors.
- Since the Internet infrastructure and Internet markets are global by nature, global approaches are needed also in the policy, governance and regulation. Regulators that impose uniquely local regulatory burdens, or more costly requirements than other regulators do, can handicap players in their national markets.

3.4.2 New Ways of Information Delivery

3.4.2.1 Overview

The business use case New Ways of Information Delivery is aimed to explore the business opportunity arising from the technical developments of the Network of

Information concept. The Network of Information concept is based on the central role that information will have in the Future Internet, as the opposite to the central role that connections have in the current Internet [10, 18].

The underlying business idea is to support the digitalisation of media (books, video, tapes, etc.) and to provide the essential functions for a comfortable access to digital media any time on the best possible device. The Network of Information approach enables one to be always in touch with the intangible part of everybody's life (i.e., information of various kinds). In fact, a seamless access will be provided everywhere, through any device, to any service and personal data.

Major benefits for end users appear to be the opportunity of a comfortable and quick optimised information delivery, coupled with an optimised presentation format that will depend on the end device and on personal favourite configuration options, enabled by the simple switching feature between different end devices.

Major benefits for providers will consist in new service opportunities, such as the possibility to provide efficient and billable ways of information delivery, the opportunity to expand offers to the whole world and to overcome today's limited service usability, and the fact that any customer at any time will always be able to reach any information object, with no concerns about broken physical links or changed location addresses. Cost reduction is also an aspect to take into consideration.

The business processes will be modified, as a consequence of the ongoing changes in value chains, starting from the shape itself: the value chain for the use case will change to a value cloud, as process flows will not be settled in a unique way, rather being adaptive. Then, as the Future Internet will provide more efficient and billable ways of information delivery as in current file sharing networks, new or improved product features will easily emerge. This is based on the assumption that the market is willing to pay for a comfortable, fast and secure information access, which will be nearly automatically configured by specialised end devices.

A challenge for the 'New Ways of Information Delivery' is about security, privacy and confidentiality issues. On the security side, a clear access ruling (one has to consider both access control and regulation on access) must be granted for stored information on any physical platform both for information providers and for information users. Another aspect of the same problem is to ensure also the end-to-end network security. In fact, the risk for an unauthorised disclosure of personal data, stored in a cache memory, increases as the number of cache memories grows. On the privacy side, protection is needed for information objects stored in the cache memories of users' devices, so that third parties can be prevented from unauthorised access to sensitive data. Spam prevention, identity certification of information providers and maintaining data integrity in any type of memory storage are further challenging problems that appear to be of great impact on societal and business aspects.

One is expecting improvement in network quality by decreasing access time and improved up/download times, therefore there should be billing opportunities for these valuable qualities.

Another valuable economic issue is 'going green': increased network efficiency will provides less retransmissions, less handovers, increased capacity, and reduced energy consumption. Hence, there is a beneficial environmental impact, due to many objects that will become intangible and will not need to be either transported or recycled any more.

3.4.2.2 Regulatory Perspectives of Security, Privacy and Confidentiality Issues

Today, the Internet is already seen as a critical infrastructure, and this will be even far truer in the future. Therefore, in the Future Internet security, privacy and confidentiality will be among the key objectives when designing the new information sharing concepts. Network and information security are vital for business transactions and for the protection of personal privacy [19–22].

For identifying issues in the Network of Information concept, several information retrieval use cases were created and analysed. The analysis was done against the generic requirements for security, privacy and trust, and resulted in the identification of the potential security, privacy and trust issues in the concept [10], which are listed below:

- Information Security—Information security is related to the requirement that the use of electronic communications networks to store information, or to gain access to information stored in the terminal equipment of a subscriber or user, is only allowed on condition that the subscriber or user concerned is provided with clear and comprehensive information on this (e.g., in accordance with EC Directive 95/46/EC [22]), inter alia about the purposes of the processing, and the subscriber is offered the right to refuse such processing by the data controller. The concern is that user information, or a part of it, may be located on different administrative domains, and the risk that information is accessed by unauthorised persons increases; an additional concern arise from the fact that user information may be located in different countries, due to different regulations.
- Network Security—Network security is related to the requirement to protect sensitive data from unauthorised access or accidental disclosure, the problem being typically divided into integrity and confidentiality. The integrity problem affects public information, and can be addressed by signatures and checksums that need to be verified, while confidentiality requires encryption. The more problematic aspect of trust in a network is related to authentication, access control and authorisation, when the first question to be checked is whether you are connected to the entity you intended, with no malicious middlemen. Copies of an information object may be located in different places of cache memories along the path between the entities that have exchanged the information object. As the number of cache memories, where information is stored, increases, the risk that information is accessed by unauthorised persons, increases as well. Also, a concern is that a middleman may be able to modify any information being delivered.
- Communication Security—The purpose of Communication Security is to ensure that information flows only between authorised end points of a communication path. This dimension deals with measures to control network traffic flows to prevent traffic diversion and interception. The system has to provide a tool that allows

the authorisation of a publisher using a third party as a Public/Private Key Manager. The Public/Private Key Manager can verify the publisher of information to a user. This is important when retrieving critical information.

- Data Integrity—Maintaining data integrity is a key requirement for the future critical infrastructure, and it is up to the publisher of information to ensure the data integrity of an information object. Maintaining the integrity of information located in different places of the network is a challenging task however, and the system shall ensure the integrity between the information objects in all of those places. When the content of an information object is changed, the changed information should be available for all users at the same time.

- User Privacy—Privacy is commonly understood as the right of individuals to control what information related to them may be collected and stored and by whom, and to whom that information may be disclosed. By extension, privacy is also associated with certain technical means (e.g., cryptography) to ensure that this information is not disclosed to any other than the intended parties, so that only the explicitly authorised parties can interpret the content exchanged among them. The 'Network of Information' concept allows the distribution of any type of information and the storing of that information in cache memories. That information could be very personal and sensitive healthcare related, for example. Also, combining information from several sources through data mining may result in sensitive findings, which could be widely distributed. Different countries have different views on user privacy.

- Confidentiality of Communications—Most commonly, privacy and confidentiality are used as the same term, but it should be noted that ITU-T Recommendation X.805 [23] differentiates privacy and data confidentiality, the former relating to the protection of the association of the identity of users and the activities performed by them (such as online purchase habits), and the latter relating to the protection against unauthorised access to data content. Encryption, access control lists, and file permissions are methods often used to provide data confidentiality. The Network of Information concept can be used for delivering different type of information, and the opportunities for listening, tapping, storage or other kinds of interception or surveillance may be numerous. Confidentiality of communications would be increased if confidential information would not be cached longer than needed to ensure the fast delivery of a data object to the next cache memory or to the end-user. Information that is to be delivered to a large number of people could be cached for a longer time period. A mechanism should be developed in the concept for differentiating between the different types of information and for allowing the different handling of information in the cache memories depending on the type of information.

- Availability of Security—The Availability of Security ensures that there is no denial of authorised access to network elements, stored information, information flows, services and applications due to network interruption. Network restoration and disaster recovery solutions are included in this category. The Future Internet will be used also for such services, where the availability of security is vitally important; the distribution of mission critical information, or the availability of sensitive information, are just examples of such services. The dependability of the critical system components has to be designed and implemented.

- Authentication—Authentication is the provision of proof that the claimed identity of an entity is true; in here, entities include not only humans, but also devices, services and applications. There are two kinds of authentication: data origin authentication and peer entity authentication. When certifying the identity of the originators and end users of information (and avoiding frauds based on identity theft), authentication is of utmost importance. The external public/private key managers can provide the origin authentication. A regulator may set rules for a party willing to act as a public/private key manager, since the public/private key manager will have an important role in ensuring the viability of the Future Internet as a critical infrastructure.

- Lawful Interception by Legal Authorities—Lawful interception is the interception of telecommunications by law enforcement agencies and intelligence services, in accordance with local law and after following due process and receiving proper authorisation from the competent authorities. Traditionally, these agencies have been interested mainly in the conversational services (phone calls), but critical illegal information exchange may also take place in data communication. That information may be distributed to a network of people (e.g., criminals) using the Network of Information concept: information is stored in different cache memories and it may be retrieved from different locations. Since both the location and user of information may move, it is very challenging for a legal authority to intercept the specified traffic.

Potential regulatory actions needed in the context of the Network of Information concept are listed in the following (as in the case of Network Virtualisation, an action may be relevant to several issues listed above):

- The system hierarchy with the root system has to be set up for the concept, and that root system shall be operated by a globally accepted authorised entity. The standards for exchanging information between the different hierarchy levels of the system have to be specified.

- The identity management for certifying the identity of the originators and end-users of information shall be allowed only by the highly trusted parties.

- The availability of the tools and processes to handle information and to delete sensitive information everywhere in the system shall be required. The system hierarchy has to be set up to initiate actions also at a higher level, and the highest level should be operated by a trusted party. Also, unnecessary copies of information shall be prevented.

- The availability of a tool, which allows a legal enforcement agency to monitor the traffic in the system, shall be required. The primary or any secondary storages of information are the natural points to allow the legal interception. A potential regulatory aspect is that the used keys for data publications shall be stored for as long periods of time as the individual data objects. Since a publisher of information uses the same private key for generating identifiers for his information, monitoring by a law enforcement agency could be based on the usage of the identifiers.

- Classification of information (high/low integrity and privacy requirements) is needed to allow different handling of different types of information. Common standards for the classification of information according to the required integrity or privacy level have to set up and used.
- National and regional regulators have to cooperate in order to ensure the security, privacy and confidentiality of the information exchange across the borders and globally.

3.4.3 Overview on Internet of Things

For the research community and different industry areas, the topic "Internet of Things" has been around since several years. Besides the aspects of devices like RFIDs or sensors, other topics are challenged, like connectivity of devices, application utilising the infrastructure components, and a changing ecosystem of industries and players for business opportunities. For the business use case evaluation, the 'thing' will be an object and, to be more precise, the focus is on physical objects (devices). So, in a more general context, the Internet of Things concept is coupled to the idea that in future every device, or even more general everything, will be connected to the network and will exchange information [10, 18].

There are a number of use cases that are already deployed, like RFID, or are under development, like different sensors in healthcare, or still under research, like autonomous driving cars or electronic sheets of paper. In general, the Internet of Things will cover all areas of life, and will not be just a computer scientist playground, which demands for more cross-industry collaboration and information exchange.

It is difficult to define requirements for Internet of Things. The different use cases are widely spread over communication systems theory. Use cases might demand real time or non-real time information distribution, one-to-one or one-to-many connections, transmission of a few bits up to megabytes of data. Other features, like physical connectivity or security demand, differ in the same way.

But, in general, the devices need connectivity with 'something else', another device, a sink, or a source 'in the cloud' of the network. The existing Internet with today's technologies enables the connectivity for the examples already mentioned. The Future Internet will improve the connectivity for thousands of network devices, e.g., via the Generic Path concept. A device is comparable to the definition of a 'Generic Path' entity as described in [13]. In addition, this connectivity must be able to transfer data among devices, sinks, and sources according to the needs and threats of this information. This transfer, or even more general the principle connection between the end points of such a transfer, is comparable to the definition of the 'Generic Path' itself [13]. Assuming that the diversification of different connectivity types, networks, servers and stakeholders of the connectivity will remain fragmented, or will even increase, the connectivity must be capable of performing different transitions or transformations along the path. A transformation is done in mediation points of the 'Generic Path' concept (see Definition 3 in [13]). Mobility or even multi-homing is increasing the complexity of the transmission of these data to a further extent.

In addition, network virtualisation might provide mechanisms to reduce the technical complexity or increase privacy support. The 'Network of Information' concept might enhance the development and operation of, e.g., sensor networks with standardised mechanisms.

From a general perspective based on the analysed use cases, one can conclude on the impact on the future in terms of business and social aspects at large. The Internet of Things translates existing information infrastructures of industries into new digital information.

Telecommunication/Internet provider will be the link between the devices containing the information and the servers hosting applications utilising the information. This plane is defined as resource plane and represents the supply chain side with the capabilities and the partners. On top of the resource plane is the service plane representing the value proposition to customers. This service plane consists of all the composable data, tags and information objects. The distribution side of the value chain will be users and/or producers (prosumers). It is most likely that a varying number of players will take over many business roles in the value chain (which still needs to be fully identified).

For a business entity itself, many changes will happen. Currently, businesses are transforming a number of support processes with the help of the Internet (e.g., by making information widely available, in a very fast way). This includes marketing, customer relationship management, logistics, financial and accounting or human resources. With the Internet of Things, core processes will be impacted, like the analysis of weather conditions in agriculture, the personal information in healthcare scenarios, or user behaviour analysis for contracts in insurance industry; this impact comes from, e.g., the use of information from sensors in real time, and its processing by adequate entities.

Impacts on the society are difficult to estimate, but, if there are economically viable developments of Internet of Things, they could be enormous. Comparatively "basic" jobs, like cashing in supermarkets, could become obsolete by autonomous, customer convenient sensor based systems. This would increase pressure not only on companies employing cashiers, but also on public policies request to analyse the impacts on an industry with several million employees in Europe. But, in general, more research with cross-industry experts is required to detail the influence.

3.4.4 Overview on Community-Oriented Applications

Today's Internet users are not only spectators, as they were until recently, but are becoming producers and suppliers of content, knowledge, connection, bandwidth, context, etc. The use case on the provision of user oriented applications deals with some of these emerging new opportunities: it enables Ad-hoc Communities

(AdHCs).These communities will have a short lifetime, they are instantiated on the fly ('ad-hoc'), and they are created for very specific purposes. The peculiar aspect is that the founder of a new community could be either a person or an object (e.g., appliances, health care machines, cars, weather observation machines, etc.). The foreseen opportunities are that the new AdHCs will possibly be born to provide a fast and secure way to share something in time-sensitive scenarios and to cope with many users needs: to save time in accessing information, to find trusted information and sources of information, to have security with information confidentiality, to reduce costs of any type, and to take care of the environment.

While related business use case built on AdHCs are also addressed in Chap. 13, the major findings are summarised in the following lines:

- Innovative business models are enabled by the opportunity for all business players to offer their contents, platforms and services to a wider audience, in a trusted environment, with a seamless and ubiquitous access.
- Competition will be very strong, as entrance barriers will be lower and almost every player can easily become an e-content provider, but virtualisation, that enables infrastructure cost reductions, will not allow very large benefits to industries making use of economy of scale.
- The overall social wealth and the environment will benefit from largely avoiding business trips, reducing companies physical premises and workers commuting, providing contents for social support and ease the life to everybody, etc.
- A new economic opportunity enabled by the AdHCs is to bill the sustaining of security, trust, reliability and accessibility for all information.
- Today, a growing trend towards self-servicing often means externalising costs onto users and putting them under stress; AdHC can be used to perform some tasks in their place, avoiding unrecoverable errors.

3.5 Conclusions

The success story of the existing Internet shows that progress has not mainly come as a response to requirements arising from the socio-economic and regulatory side, but technical progress has driven adaptation of, e.g., economic models and regulatory rules.

As a consequence, there is the need to carefully monitor and assess the outcome from new technical developments, which may initiate new services and applications, and thus will have impact on our society and economy. Due to the mutual influence of technical and non-technical driving forces, all these interactions need to be done in iterative steps. Drawing the appropriate conclusions from the interdependence of the non-technical driving forces with technical issues is the key for the deployment of Future Internet innovations. The main findings are outlined below.

Migration towards information-centric networks is a key issue. Users are mainly interested in using services and accessing information, not to be aware of the location of the service realisation or the information. The faster, the easier and the safer any type of information will be accessible, the more the Future Internet will have a disruptive social impact on a rapid spread of Internet services through all social, age and educational strata.

There is the need to **design new, advanced connectivity services to be used by humans or 'things'**. The Internet of Things will be an important part of the infrastructure of the Future Internet. The presence of thousands of different network devices will result in a big variety of connectivity requirements and in an enormous bunch of applications.

The opportunity to have a widespread Internet of Things will open new life scenarios where 'things' will take over the responsibility to perform tasks that are today in the hands of human beings. For example, home appliances going out of order take by themselves the responsibility to call the proper customer centre, cars that need to be repaired connect to the garage to order assistance, or health devices (e.g., pacemakers) contact the medical aid in case of sudden changes in the bearer health conditions.

There will be the **creation and deployment of new types of networks via virtualisation**. The growing complexity, diversity and heterogeneity of networks are major issues in network operation and maintenance. Future networks, employing also self-management capabilities, will significantly reduce networking OPEX and CAPEX. Network providers will be able to choose between either to invest in dedicated new physical network resources or to act as virtual network providers and use the physical resources of other providers. Similarly, the need for local customer service and local assistance in physical networks will be partially substituted by the growing number of delocalised software companies that can be based everywhere. The social impact of remote and delocalised business will consist of a reduced migration into cities, thus, preventing all related social problems.

Security and privacy will be improved. In the Future Internet, security, privacy and confidentiality shall be among the key objectives when designing new information sharing concepts. Network security and information are vital for business transactions and for the protection of personal privacy. The confidentiality of communication and related data traffic shall be ensured by prohibiting the listening, tapping, storage or other kinds of interception or surveillance by persons other than the ones involved in the communication without the consent of the users concerned.

The management of information privacy and related responsibilities need to be clear. It will be a major challenge from the legal perspective, on the one hand, to ensure privacy and security of data, and on the other, to keep the Future Internet an open platform for business and administrative applications, entertainment, information exchange, etc.

The great challenge for secure networking applications will be due to the huge number of objects (several orders of magnitude greater than the number of today's devices or connections) they will have to handle, in both physical and virtual networks.

Interconnectivity, Interoperability and Standards need to be addressed. In the IP world, interconnection has been normally implemented on voluntary agreements between IP Service Providers. These freely negotiated arrangements have resulted in a richly interconnected Internet and have not depended on regulatory obligations. The Future Internet, which will be more complex with several different technical and administrative domains (e.g., virtual networks) and carrying new types of traffic, raises concerns on whether the old approaches are sufficient in the future. Basic principles should define the set of information that networks have to exchange in order to make the Future Internet work correctly.

The Future Internet ecosystem will also comprise of many new players, which are partly rivals and partly collaborators in their offerings to the customers. Additionally, the Future Internet will remain split into many administrative and legal domains. Both aspects demand for widely accepted standards for interfaces and rules for interoperability. Focusing on the interfaces ensures the interoperability while preserving, e.g., network operators' freedom to use customised solutions inside their networks. Standards shall be robust in terms of completeness (complete technical disclosure of APIs, protocols and formats), control (fair and transparent multilateral governance), cost (fair, reasonable and non-discriminatory licensing of essential IPR), and compliance (adherence to standards and industry specifications).

The debate on the **Network Neutrality** is a business related discussion about charging schemes, and a technical issue concerning the need for traffic prioritisation related to the potential network congestion. Transparent and competitive broadband markets should ensure that service providers continue to be able to provide innovative services across networks. However, the interests of service providers and network operators need not to diverge: the Internet can be characterised as a two-sided market, where the success of service providers depends on the adequate and high quality access networks and vice versa. In the end, all players have to ensure that this symbiotic ecosystem is not destroyed.

The regulatory framework for electronic communication networks shall allow and ensure the non-discriminatory network access for the service providers, and also enable incentives for operators to invest in and maintain high-quality and high-speed networks. The ongoing discussion on Network Neutrality has to cover also new concepts like Virtual Networks.

Harmonisation of rules across borders is a need. Since the Future Internet will be a combination of thousands of networks, and since a single network may span across countries and regional borders, the variability of rules across borders will represent a hurdle to deploy seamless end-to-end services in multi-provider environments. The end-to-end principle of a service over the network and country boundaries simply cannot be implemented without commonly accepted rules.

Regulators that impose uniquely local regulatory burdens or more costly requirements than other regulators do, can handicap players in their national markets. It may happen that the legal basis of operations will be moved to countries where rules are missing. There are regional organisations, like the European Regulators Group (ERG) that are trying to harmonise the rules on a regional area, but the collaboration between regions, however, may be challenging.

References

1. P. Mähönen, D. Trossen, D. Papadimitriou, G. Polyzos, D. Kennedy, The Future Networked Society, a white paper from the EIFFEL Think Tank, ICT-EIFFEL Project, White Paper Report (Dec. 2006), http://www.fp7-eiffel.eu
2. T.-R. Banniza, A.-M. Biraghi, L. Correia, T. Monath, M. Kind, J. Salo, D. Sebastiao, K. Wuenstel, First project-wide assessment on non-technical drivers, ICT-4WARD Project, Deliverable D-1.1 (May 2009), http://www.4ward-project.eu/index.php?s=Deliverables
3. E. Kountz, P. Wannemacher, B. Ensor, C. Tincher, The recession's impact on US consumers' financial behavior, Forrester Research (Feb. 2009), http://www.forrester.com/rb/Research/recessions_impact_on_us_consumers_financial_behavior/q/id/53556/t/2
4. United Nations, World Population Ageing 2009, Department of Economic and Social Affairs—Population Division (2009), http://www.un.org/esa/population/publications/WPA2009/WPA2009_WorkingPaper.pdf
5. A. Henten, R. Samarajiva, W.H. Melody, Designing next generation telecom regulation: ICT convergence or multisector utility?, WDR, Report on the Dialogue Theme 2002 (Jan. 2003)
6. European Commission, Decision establishing the European Regulators Group for Electronic Communications Networks and Services, Decision 2002/627/EC (29 July 2002), http://www.erg.eu.int/doc/legislation/erg_establish_decision_en.pdf
7. ITU, Trends in Telecommunication Reform: The Road to NGN, 8th edn., Geneva, Switzerland (Sep. 2007), http://www.itu.int/publ/D-REG-TTR.9-2007/en
8. H. Intven, M. Tétrault (eds.), *Telecommunications Regulation Handbook* (infoDev, Washington, DC, 2000). http://www.infodev.org/en/publication.22.html
9. Social and economic factors shaping the future of the Internet: Proposed issues list, in *Proc. of NSF/OECD Workshop*, Washington, D.C., USA, Jan. 2007, http://www.oecd.org/dataoecd/35/24/37966708.pdf
10. M. Soellner (ed.), Technical requirements, ICT-4WARD Project, Deliverable D-2.1 (Apr. 2009), http://www.4ward-project.eu/index.php?s=Deliverables
11. http://trilogy-project.org
12. OECD resources on policy issues related to Internet governance, http://www.oecd.org/site/0,3407,en_21571361_34590630_1_1_1_1_1,00.html
13. F. Guillemin (ed.), Architecture of a generic path, ICT-4WARD Project, Deliverable D-5.1 (Jan. 2009), http://www.4ward-project.eu/index.php?s=Deliverables
14. S. Baucke, C. Görg (eds.), Virtualisation approach: Concept, ICT-4WARD Project, Deliverable D-3.1.1 (Sep. 2009), http://www.4ward-project.eu/index.php?s=Deliverables
15. U. Drepper, The cost of virtualisation, ACM Queue **6**(1) (Jan./Feb. 2008)
16. M.E. Porter, The five competitive forces that shape strategy, Harv. Bus. Rev. (Jan. 2008), http://hbr.org/2008/01/the-five-competitive-forces-that-shape-strategy/ar/1
17. P. Reynolds, B. Mitchell, P. Paterson, M. Dodd, A. Jung, P. Waters, R. Nicholls, E. Ball, Economic study on IP interworking, White Paper, CRA International and Gilbert+Tobin (Feb. 2007), http://www.gsmworld.com/documents/ip_intercon_sum.pdf
18. B. Ohlman (ed.), First network of information architecture description, ICT-4WARD Project, Deliverable D-6.1 (Apr. 2009), http://www.4ward-project.eu/index.php?s=Deliverables
19. ITU-T, Security in Telecommunications and Information Technology, Geneva, Switzerland (Dec. 2003), http://www.itu.int/itudoc/itu-t/85097.pdf
20. European Commission, Directive concerning the processing of personal data and the protection of privacy in the electronic communications sector (Directive on privacy and electronic communications), Directive 2002/58/EC (12 July 2002), http://eur-lex.europa.eu/pri/en/oj/dat/2002/l_201/l_20120020731en00370047.pdf)
21. European Commission, Communication on strategy for a secure information society—Dialogue, partnership and empowerment, COM(2006) 251 (30 May 2006), http://ec.europa.eu/information_society/doc/com2006251.pdf

22. European Commission, Directive on the protection of individuals with regard to the processing of personal data and on the free movement of such data, Directive 95/46/EC (24 Oct. 1995), http://ec.europa.eu/justice_home/fsj/privacy/docs/95-46-ce/dir1995-46_part1_en.pdf
23. ITU-T, Security architecture for systems providing end-to-end communications, Recommendation X.805, Geneva, Switzerland (Sep. 2003), http://www.itu.int/rec/T-REC-X. 805-200310-I/en

Chapter 4
Network Design

Susana Perez Sanchez and Roland Bless

Abstract A proposal is presented for a possible Architecture Framework covering certain demands of the Future Internet. Some concepts, terms and the basic constructs are defined, in order to model Network Architectures. This Architecture Framework provides two levels of views on network architectures: the macroscopic view, mainly focusing on structuring the network at a higher level of abstraction, and introducing the concept of Strata; the microscopic view, concentrating more on the functions needed in the network nodes, their selection and composition to Netlets. Functional Blocks are presented as the common points between the two views of the architecture. The Component Based Architecture constructs and principles are used as the basis to provide reusable frameworks that minimise the design and development times of new network architectures.

4.1 Introduction

The Internet has evolved to a global, indispensable network infrastructure. Despite its great success and all innovations that it already brought, it also represents a stalemate with respect to new network architectures. The Internet Protocol IP is an invariant on which the current Internet architecture is built. The Transmission Control Protocol TCP, Domain Name System DNS, or Border Gateway Protocol BGP for inter-domain routing constitute similar invariants. Changing or replacing these protocols is hard. Integrating and deploying new protocols is likewise difficult. Well-known examples are the deployment problems of IPv6 or IP multicast, which require modification or an upgrade of the installed network nodes.

S. Perez Sanchez (✉)
Tecnalia-Robotiker, Zamudio (Vizcaya), Spain

R. Bless
Karlsruhe Institute of Technology, Karlsruhe, Germany

L.M. Correia et al. (eds.), *Architecture and Design for the Future Internet*,
Signals and Communication Technology,
DOI 10.1007/978-90-481-9346-2_4, © Springer Science+Business Media B.V. 2011

Contrastingly, in the context of the Future Internet it is envisioned that different networks constructed following different architectures can co-exist and share a common infrastructure. These network architectures can be specifically tailored to particular user or application requirements and, furthermore, can take into account the characteristics of the available networking resources. One of the main objectives of the 4WARD project is the development of an *Architecture Framework* to represent and design future network architectures within families of interoperable networks.

In order to manage the complexity of communication systems, different levels of abstraction are introduced. Typically, the concept of layering is applied to communication systems with the ISO/OSI reference model as the most famous representative. However, lately many arguments have been raised against such a strictly layered model which basically treats layers as black boxes. These arguments, for example, refer to the fact that cross-layer information is needed in order to design efficient and effective communication systems. Moreover, such models might be limited with respect to communication systems with specific requirements, for example, sensor networks, the Internet of Things, or the Network of Information. One goal of 4WARD is to develop an Architecture Framework that holds for such a variety of networks. Therefore, a suited structuring of the network architecture is needed.

The 4WARD tenet "Let 1000 networks bloom" expresses the objective to let a variety of such specifically created and specifically tailored Network Architectures co-exist on the same network infrastructure. *Virtual Networks* (*VNets*) represent a possibility to overcome the initially mentioned Internet ossification that is caused by the previously mentioned invariants like IP. If future networks support VNets, deployment of new and alternative network architectures, which may be designed using the above mentioned Architecture Framework, will become easier. VNets basically consist of virtual nodes and virtual links between those nodes. For users of the virtual network, the virtual network resources look like ordinary network resources.

On the one hand Network Virtualisation can be seen as an enabling technology for introducing and deploying new network technologies more easily. For instance, new protocols of different network architectures on top of virtual link layer topologies can be deployed. In contrast to currently used virtual network technologies, such virtual networks offer the ability to run arbitrary platforms—not necessarily IP-based ones—inside the virtual nodes. Thus, Network Virtualisation offers a possibility to smoothly test, debug, and roll out new network architectures in parallel [4], including the interconnection and interoperability among them, which is a specific requirement for the Architecture Framework definition. Consequently, Network Virtualisation provides an evolutionary path towards innovative new network architectures.

On the other hand, Network Virtualisation also provides more flexibility for Infrastructure Providers (InP), customers, and users of such virtual networks. Similar to host virtualisation that allows moving virtual hosts between different physical hosts, Network Virtualisation allows moving virtual nodes and virtual links among physical nodes as well as expanding or shrinking the virtual topology on demand.

An advantage for the operator of a virtual network is the ability of virtual networks to span multiple InP domains without the operator having to directly deal with interprovider issues in his virtual network.[1] But these advantages all come at the cost of a higher management effort due to another level of abstraction: instead of managing physical resources directly, virtual resources have to be managed on top of physical resources. Therefore, Network Virtualisation approaches need to support efficient management and control mechanisms.

Given the variety of different network architectures that may reside in separate virtual networks, design support and guidance are needed when creating new communication network architectures. In order to cope with this variety and complexity of communication networks, two different and complementary views on network architectures are provided: a macroscopic and a microscopic view:

- The macroscopic view is more related to an overall structuring of the network design at a high level of abstraction in terms of so-called strata. The strata concept states a flexible way to layer the services of the network that can enable the usage of information across different layers.
- The microscopic view deals with the functionalities needed within the network architectures and their composition to so-called Netlets in order to fulfil desired requirements. In this way, the Node Architecture hosts various Netlets of the same or different families of network architectures.

Besides these architectural entities (Strata, Netlets, and Node Architecture), a Design Process to guide the Network Architect while designing new network architectures has been defined. After finishing the design of such new network architectures VNets can be used to facilitate their roll-out and actual deployment.

The remainder of this chapter presents a proposal for an Architecture Framework, guided by the description of its main basic constructs, and the description of an appropriate Design Process. Then a Virtualisation Framework is introduced, as a means to deploy different network architectures in a flexible and isolated way, specifying the VNet lifecycle, an architecture for the substrate host node, and the end user attachment to the VNet.

4.2 Architecture Framework and Its Basic Constructs

Figure 4.1 shows the 4WARD Architecture Framework, and its main features. These characteristics are briefly outlined here and explained in more detail within the next subsections.

According to the basic constructs represented in Fig. 4.1, a network consists of the vertical strata plus a set of horizontal strata. A specific network architecture would then be specified by defining a specific set of vertical and horizontal strata.

---

[1] For the interconnection of different virtual networks, the virtual network operators involved would play similar roles to the case of current physical network interconnection.

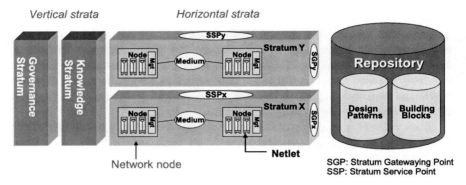

Fig. 4.1 Sketch of the 4WARD Architecture Framework

In particular, the Governance Stratum must be present in any network. The functionalities of these Strata are distributed across the Nodes. Main difference between Vertical and Horizontal Strata resides in their nature, or main function within the network: horizontal strata provide different levels of transmission capabilities, communication among different nodes that form them; however, vertical strata have the responsibility of managing and monitoring the different aspects present within the network, and taking decisions on which action to take if that is required.

As shown in Fig. 4.1, the following main components are considered in this framework:

- In the mid part of the figure, the Horizontal **Stratum** is composed by a set of **Nodes** that are connected through a **medium**. This entity encapsulates a set of functionalities that are distributed through the nodes. These functionalities are provided to other strata through two well known interfaces (that can be also distributed across the nodes): the SSP (Stratum Service Point) and the SGP (Stratum Gateway Point), which are described later.

  In Fig. 4.1, the Horizontal Strata explicitly represents the resources and capabilities required for the communication across networks. That means that these strata will encapsulate the data plane capabilities of the network.

- The Horizontal Stratum is composed of **Nodes**. Functionalities within a node are realised with Building Blocks/Functional Blocks and logically grouped into so-called **Netlets**. The Netlets can be considered as containers that provide a certain service. They consist of functionalities that are needed in order to provide end-to-end services, or in-network services such as routing. The collection of all the functionalities leads to the definition of protocols provided by a Netlet. The Netlets contain protocols and these protocols constitute the medium for the Strata they belong to. Therefore, inside the same Netlet there could be functionalities that are related to different Strata.

- The left side of the figure represents the set of functionalities related to Network management and control by means of the Vertical **Governance** and **Knowledge** strata. These capabilities can be implemented by means of a set of dedicated nodes (that can be organised as another stratum whose main functionalities are related to network organisation, context awareness, etc.) or distributed across other nodes that are part of the horizontal strata defined before. Both the Governance and Knowledge strata will provide a well known set of interfaces (that can be also represented by SSP and SGP reference interfaces) to other strata that could allow, e.g. the negotiation of the Interconnection between two different networks.

- Finally, in the right side of Fig. 4.1, there is a *Repository* that represents the set of *Building Blocks* and *Design Patterns* to be used to support the design process. In order to build this Design Repository and to support the definition of the different functionalities, Software Engineering methodologies can also be integrated. So, for example, using Component Based Architectures (CBA) has proven to be an effective approach to address the different requirements.

The key element to link both the Stratum and Netlet concepts is the Functional Block (FB). It is defined as a sequence of instructions that realises certain functionality, therefore protocols and other functions can be built with them. Different grouping of FBs results in Netlets or in Strata, but the FB is the minimum individual entity for both.

The concept of Netlet is an extended protocol stack, which contains different Functional Blocks that are traditionally layered or combined in an arbitrary way. In today's OSI-based network, a Netlet is a layered set of FBs, interacting via Interfaces, and logically grouped together, to provide network services to a node and its applications and users.

The Functional Blocks can also be grouped in another way: similar FBs maintaining today's OSI-based protocol can be logically grouped in the form of a Stratum (e.g. Horizontal Stratum). In summary, a Netlet is a set of Functional Blocks to provide a service, and a Stratum is a set of Functional Blocks distributed over a number of nodes to provide a particular functionality.

Considering this basis, the following properties apply to the relationship among Strata and Netlets:

- Each of the Functional Blocks of a Netlet is a member of a Horizontal or Vertical Stratum accomplishing the corresponding functionality in a network.
- A Netlet in a node may be considered as a set of Functional Blocks of different types to make a node operate in a network and to offer services to the applications and users in the nodes.
- A Stratum may be considered as a set of Functional Blocks from the same or related types distributed over a set of nodes to accomplish the corresponding functionality.
- A Functional Block is the only entity shared between a Netlet and a Stratum.

Figure 4.2 illustrates how the Functional Blocks form the intersections between both Netlets and Strata. In particular, it can be seen that inside a physical node (using, e.g. the Node Architecture, explained in following subsections) different types of strata (including also vertical and horizontal ones) are represented by their functional blocks implemented inside Netlets. In the figure, FBs are represented by the intersectional points among strata and Netlets.

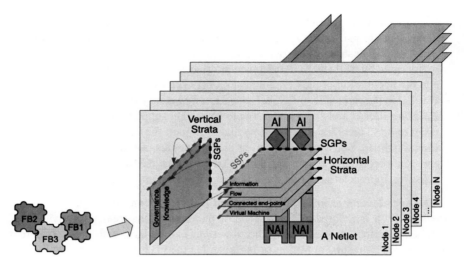

Fig. 4.2 Relationship between Netlets and Strata

The CBA (Component Based Architecture) represents the set of concepts that could be used to implement the proposed architectural components. These architectural components are used to support the implementation of the Strata.

4.3 Network Design: Strata

The macroscopic view of the Architectural Framework goes beyond the concept of the OSI layering and supports the design of network architectures following a modular and flexible "black box" approach; that means that specific functions can be implemented over any SW or HW platform thanks to the specification of the proper external interfaces. The internals of each specific implementation (e.g. inside a specific network), are encapsulated and hidden by the external interfaces. Another manner to deploy and implement strata based architectures is to create a Virtual Network as an overlay above a set of physical resources. In this way, the Vertical Strata are involved in the instantiation and management of VNets, since they bring together functionalities associated to the orchestration of the VNet Operator architecture within the virtual network. Management of VNets will be specifically addressed in following sections.

A *stratum* is by definition a distributed function. The distributed function is modeled as being distributed across a set of *logical nodes*[2] (and those logical nodes

---

[2]*Logical Node* is not the same as *Virtual Node* (as defined in Network Virtualisation). It refers to the *distributed* feature.

Fig. 4.3 A stratum, its
internal structure and
interfaces

are distributed across a set of physical nodes), where data can be stored and processed. The piece of functionality residing in each logical node communicates with pieces of functionalities residing in other nodes through the **medium** of the stratum.

The **Stratum Service Point** (**SSP**) represents the set of interfaces that offer the services provided by each stratum. An SSP is further decomposed into one or more interfaces. The way this partitioning is done depends on modularity aspects like security, manageability, cross-layer support, etc. The SSP also models the encapsulation of the functionalities being distributed across the set of nodes (and communicating via the medium), i.e. the users (other strata) of the (end-to-end) services offered via an SSP do not need to be aware of how the stratum with nodes and medium has been defined (thus supporting the black box principle).

The **Stratum Gateway Point** (**SGP**) provides access to other strata being of the same or similar type (i.e. having a common point of origin regarding its specification), but independently realising functionalities in different networks. SGPs are points where strata may interoperate if necessary. An SGP is further decomposed into one or more Interfaces.

A stratum, with nodes, medium, SSP, and SGP is depicted in Fig. 4.3. The logical nodes (N) and the medium can be referred to as the internal structure of the stratum, which is important for its instantiation and deployment. The SSP and the SGP can be referred to as the external interfaces of the stratum, which encapsulate and hide its internal structure. It shall then be further noted that the internal structure of a stratum may change without any need to change the specification of the SSP or the SGP. This allows different designs and implementations of the internal structure of the stratum.

Different operations can be associated to the stratum concept.

- First of all, a stratum is instantiated and deployed and further managed following concepts of SON[3] (Service Oriented Networks): the process to instantiate the pieces of the functionalities in the different logical nodes can be orchestrated in such a way that e.g. the network creation process can be optimised in terms of operational costs.
- Secondly, strata can compose among themselves in such a way that the following composition types are identified:

---

[3]Service Oriented Network (SON): designed following a flexible set of principles during the system development and integration phases. Services comprise unassociated, loosely coupled units of functionality that have no calls to each other embedded in them. Functionalities use protocols to describe how services pass and parse messages, using description metadata.

    o **Service Composition:** strata of different nature can self-compose in order to provide a specific service entity.

    o **Network Composition:** strata of the same nature self-compose in order to extend a specific service entity across network boundaries. In this case, the involved strata negotiate and implement the agreement through the SGP.

All these composition operations can be modeled through the defined strata operations (concatenation, merge, slice and aggregation) which are described in [1].

As commented before, the stratum represents a distributed functionality. It is obvious that the nature of network functionalities can differ. In this sense, two major types of strata are identified:

- The *Vertical Strata* whose main goal is to assist in management of the network. The *Governance Stratum* aims to check that a proper set of horizontal strata are instantiated and properly configured, via policies and with the information about the current status of the network. The *Knowledge Stratum* provides information to other strata about the topology of the network, current resource status, context information etc. It also monitors the status of the network continuously by collecting, storing and processing status information from the other horizontal strata and discovers new capabilities in his or other domains.

- The *Horizontal Strata* are composed by a set of strata which basically provides the resources and capabilities for communication across networks. The *Machine Stratum* provides the underlying processing and transmission capabilities to other strata (it can be constituted by physical or virtual resources). The *Connected Endpoints Stratum* provides the "road" infrastructure for communications. The *Flow Stratum* provides the capabilities for the transfer of data across networks, and finally the *Information Stratum* handles the management of data objects in networks.

Additionally, several generic properties and functionalities at different levels of abstraction can be defined and inherited by the aforementioned strata. Examples on such kind are Security, Mobility, QoS, and Self-Management properties. This in turn can be used in order to build and create vertical and horizontal strata that need to meet a specific set of requirements. Therefore, these functionalities and properties are part of the Design Repository and will be used by the Network Architect to pick up specific solutions and guidelines to meet the requirements.

4.4 Node Design: Node Architecture

To accommodate new network architectures and protocols quickly, and to access multiple networks of different kinds at the same time, a generic framework for Netlets is proposed: the Node Architecture. The Node Architecture concept provides considerations regarding the microscopic view of the Architecture Framework (see detailed description in [11]), especially with respect to the selection of a Netlet depending on communication requirements imposed by the application or user.

Fig. 4.4 Outline of the
End-Node Architecture

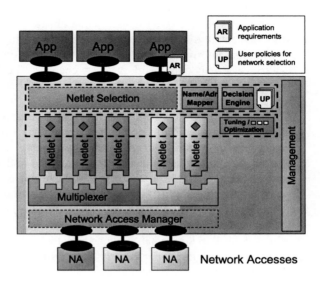

Figure 4.4 shows the outline of the Node Architecture. It describes a possible design of end-nodes supporting a multitude of current and future protocol stacks. These protocol stacks are encapsulated in *Netlets*. Within the Node Architecture, Netlets are considered as black-boxes. Thus, they can be designed, for instance, manually by just writing code, following a protocol composition approach based on functional blocks, and/or using a design tool with partial code generation.

For a virtualised substrate node, this picture could be replicated, as to. so many instances of a Node Architecture as needed (multiple virtual nodes can reside in a single physical node) (cf. Fig. 12.5). Each virtual node can follow a different combination of Netlets and different design, since it may belong to a different VNet, with a different nature. More details about the support for virtualisation are explained in a dedicated section of this chapter.

When running multiple Netlets in the same network, they need a basic common understanding of, e.g., the data unit format or addressing scheme. The *Netlet Multiplexer* uses this basic set of network invariants to (de-)multiplex data streams. Since those invariants may differ from network to network (e.g., sensor networks, backbone networks, delay tolerant networks etc.), multiple Multiplexers are allowed to run in parallel.

In order to communicate through the Node Architecture, applications need to define their communication requirements. Based on these requirements and the requirements defined by user and/or administrator policies, the *Netlet Selection* component chooses an appropriate Netlet, taking the underlying network properties into account. This selection process is described in detail in [12]. The Name/Address Mapper is a generic component that tries to resolve the names given by the application for the different architectures. This can be seen as a pre-selection of Netlets that are actually able to resolve the given name.

The *Network Access* (NA) is used as an abstraction of the connectivity to the network. Essentially, it is similar to today's network interface, but it may provide more elaborate information about the underlying physical or virtual network, and it may trigger events on network property changes. The *Network Access Manager* is responsible for mapping available Network Accesses to the multiplexers of the different architectures.

Since changes of the network properties might occur during an ongoing communication, adaptations of Netlet configuration parameters might be necessary. This is generally handled by the Tuning/Optimisation Agent.

Certain functionalities defined in the horizontal and vertical Strata are implemented as functional blocks within the Node Architecture, either directly (if the scope of the functionality is limited to the local node) or as part of Netlets (if the functionality is distributed in nature). Typical management and governance tasks defined in vertical Strata are implemented as part of so-called Control Netlets. Functionalities defined in horizontal Strata and related to data transport are part of functional blocks found in regular Netlets. Depending on what functional blocks are implemented within a Netlet, it may belong to multiple Strata at the same time (see Fig. 4.2).

The main advantages of the Node Architecture lie in the dynamic coupling of applications and network protocols. In contrast to today's architectures, this enables protocol agility. Protocol agility allows to pick dynamically the best suited protocol for a given communication association of an application. This means that every time an application starts to communicate, the communication protocol which is most suitable at that point in time will be chosen [12].

Protocol agility is important because it allows easy replacement of protocols, as needed if for example protocols are not suitable for new communication requirements, security problems with protocols are being discovered, or if new research results should be easily incorporated into an existing network. When, for example, examining the current TLS (Transport Layer Security) protocol, it becomes clear that the agility of the cryptographic algorithms is an important part of this security protocol. However, it cannot solve the recently discovered security problems of TLS [5]. Using protocol agility, however, such problems could be easily overcome by using a different security protocol.

The protocol agility of the Node Architecture also allows running many different solutions for creating communication protocols on a single node in parallel, as long as all protocol candidates can be contained in Netlets. This allows for essential plurality of communication protocols and for different solutions to compete in supplying the best communication protocol.

4.5 Component Design: Component Based Architecture

The main challenge for software developers today is to cope with complexity and to adapt quickly to changes in design and/or implementation of functionalities, protocols, services, etc. Component Based Architecture (CBA) is an approach that is used to address these demands and is core to the Component Based Software Engineering (CBSE) [2].

CBSE has become recognised as a new sub discipline of software engineering, the major goals of which are:

- To provide support for the development of systems as assemblies of components.
- To support the development of components as reusable entities.
- To facilitate the maintenance and upgrading of systems by customising or replacing their components.

CBA plays the role of defining the components, their relationships and their functionalities within the Design Repository of the architectural framework.

A component encapsulates its constituent features as the unit of independent deployment. Thus a component needs to be concretely separated from its environment and other components. As a unit of deployment, a component cannot be partially deployed. In this context a third party cannot be expected to have access to the construction details of all the components involved.

The component specifications are defined by the contracts attached to each component. The contract provides meta-data and semantics related to the component interface and functionalities. These contract specifications of available components are stored within the design repository. The component contract forms the basis for interoperability and composition. The contract provides the formal mechanism by which interoperability between components can be measured. This interoperability then becomes the basis for composition as only components that are interoperable can be composed. The contract meta-data may also be used to construct ontologies of contracts and associated components. These ontologies when linked to functionalities such as QoS or mobility can be used to guide a designer starting from a functional requirement towards possible contracts that fulfil the requirement and, ultimately, associated components.

At the higher abstraction levels of Strata and Netlet, CBA is providing the functional blocks (components) and units of interoperability (contracts) that can be used during the composition process. The CBA level is providing a link to development and deployment. The Strata itself is a collection of components cooperating to provide the required overall function of the Strata.

In CBSE, related products and systems can be assembled from pre-built components. These reusable components can take a variety of forms, from existing software libraries, to free-standing commercial, off-the-shelf products (COTS) or open-source software (OSS), to entire software architectures and their components. CBSE offers many advantages, such as (1) increased software reuse, (2) a shortened product development time, (3) reductions in total costs, and (4) fast access to new technologies, since new software components can be purchased instead of developed in-house.

4.6 Design Process

The support of rapid prototyping of new network architectures can be seen as one of the goals of the *Design Process* which is described in this section.

The process that is proposed does not aim to solve all the problems that appear during the design of new architectures but it presents a general structure that could help to apply general software engineering principles to the design of network architectures, thanks to the specification of a process that starts from the business specification and that finishes with the specification of the components that could be quickly implemented by reusing some previous prototypes.

How this process is adopted will depend on the different roles of actors involved. So, e.g. there will be specific phases of the process that can be covered by different organisations depending on their role on the value chain (i.e. operators will usually focus more on the specification of detailed technical requirements while vendors will usually focus on the development of components able to satisfy the requirements). It should be noticed that this approach is not just proposing a software development process; in fact, a network design process that can take advantage of the software development principles is proposed.

4.7 Phases of the Design Process

The Design Process represents the workflow from a business idea as a starting point to the design of the suitable models for the network architecture. These models basically constitute the blueprint of the network design which can then be used to implement, build, and deploy the network components.

According to Fig. 4.5, the following phases are considered in the Design Process:

1. *Requirements Analysis*: starting from the business idea and the business requirements, the goal of this step is to carefully analyse those requirements, decompose them and being able to identify, at least, the high level functionalities (e.g., whether QoS enforcement will be required for implementing a specific service, or routing protocols able to support highly dynamic topologies are needed) that should be implemented for the architecture to be designed.

   The output of this phase is mainly the identification of the strata (characterised by their own functionalities), a first draft of the main network components (physical nodes) and the specification of the requirements of the architecture to be developed. If the initial set of requirements is related to the addition of a new feature in a working network, the proposed new technical requirements must consider migration aspects and the current existing network topology.

2. *Abstract Service Design*: the main goal of this phase is to consider the requirements and the high level functionalities derived from these requirements in order to identify and define the specific functionalities and how they can be composed. The result of this design phase is the specification of the Netlets and the Strata that constitute the architecture. They represent how the different functionalities must be composed and distributed across different nodes in the network in order to implement the specified functionalities.

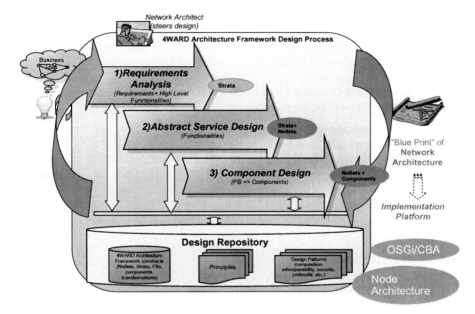

Fig. 4.5 Design Process

3. *Component Design*: it focuses on the detailed specification and composition of the Functional Blocks (FBs) used to implement the specific functionality. This includes the specification of the interfaces, properties, and requirements/prerequisites of the FBs. The output of this phase is the detailed design of the Netlets and software components that can be implemented in a specific platform after the Design Process.

A close relationship of the Requirement Analysis with the Abstract Service Design and the Component Design phases is expected. In particular, thanks to the feedback provided by the latter, the requirements can be specified iteratively with more detail. Each iteration might reveal the need to change existing or add new high level functionalities.

These phases constitute the design process resulting in a blueprint that contains: (i) the detailed requirements: mainly built during the Requirements Analysis phase that is fed with the input from other phases; (ii) the specification of the functionalities to be designed and how they should be composed in order to meet the requirements; (iii) the specification of the Strata and Netlets as the architectural elements that represent the architecture at both macroscopic and microscopic level; and (iv) Components and Netlets designs which represent the detailed specification that could be easily implemented after the Design Process.

After the Design Process, a first implementation (a prototype) of an architecture should start considering specific deployment platforms (which at the end will depend on the framework used by the stakeholder in charge of this). Two main implementation specific models are proposed; they can be also selected according to the Network Architect experience:

- The Node Architecture to instantiate the Netlets, which are also a unit of deployment.
- The CBA platform is more related to the implementation of the components and its further deployment inside the nodes.

Each implementation is complementary and the final decision will depend on the knowledge of the specific stakeholder in charge of this step.

As shown in Fig. 4.5, the design repository plays a central role in the process. This repository aims to increase the reuse of architectural constructs and store the expertise and knowledge of the designing architect. It is described in more detail in the following subsection.

4.8 Design Repository

The Design Repository contains pre-built architectural constructs (Strata, existing Netlets, components, functional blocks), proven architectural design patterns on service composition, interoperability, security, etc. and the design principles to be followed. This repository stores the results of the ongoing process in order to allow its future reusability.

Reuse has been typically opportunistic in nature, where one entity (let us state a node, an architect, a developer, a service, etc.) was able to take advantage of the efforts of another. A paradigm shift is needed from current network engineering and development practices to an engineering process in which network artefact[4] reuse is institutionalised and becomes an inseparable part of the network development process. Reuse should be systematic, driven by a demand for network artefacts identified as a result of domain analysis and architecture development. Reuse needs to be treated as an integral part of engineering and acquisition activities. In the case of the design repository presented, a network artefact can mean a reference or sample architecture, design patterns, hardware elements and software components.

An effective collection of such artefacts will guide reuse activities to avoid duplication of effort, impose necessary standardisation, and ensure repository population (user demand-driven). The Design Repository should support network artefact reuse by helping network architects locate, comprehend, and modify artefacts. At a high level, it is composed of three parts: a repository that contains artefacts, an indexing and retrieval mechanism, and an interface for user interaction. Such a repository would include: a user GUI, a standard artefact description framework (e.g. artefact purpose, functional description, certification level, environmental constraints, historical result of usage, legal restrictions, etc.), an effective artefact classification scheme.

---

[4]Network artefact represents any network element (Netlets, Strata, a specific Functional Block, a network itself) that can be designed.

The Design Repository should provide as much automated support for the network architect as possible on identification, comparison, evaluation, best practices, and retrieval of similar network artefacts. The repository must have a way to classify artefacts so that the users can quickly find what is wanted. A standard artefact description framework helps ease the process of comparison and comprehension of similar components. It would include data such as relative metrics for reusability, reliability, maintainability, scalability and portability. Inclusion of artefact documentation provides additional information to help the network architect choose an artefact.

In order to truly enable reuse and continually improve the Design Repository, its capabilities and procedures will need to be integrated within the system development and acquisition business process. It also should cover identification and support for specific requirements (e.g., security, mobility, and management) and reusable artefacts that provide these functionalities. The Design Repository should include as well tools for intercommunication and interoperability among diverse or distributed repository systems. These requirements can only be resolved through the combination of developing new technologies, standard procedures, and evolution or revision of existing policies. The same can be applied either for physical or virtual network design.

Network artefact retrieval is a fundamental issue for reuse. The retrieval process involves finding an artefact matching the desired functionality or requirements and making sure that the artefact satisfies the required non-functional properties (i.e. timing, resource constraints in the case of a network element or software component). There are many approaches that can be used to retrieve the relevant artefacts which include classic approaches (keyword, browsing), facet approach (groups of related terms in a subject area), AI approaches, Ontology-based approach, specification based approaches and automated retrieval (successive search filters).

It is important to highlight that the Design Repository should not be a static element for the Network Architect. It must be an element able to incorporate the new developments/designs done during any new design. In particular, the specification of new design patterns will be a continuous process.

The network artefacts inside the Design Repository are organised in a set of sub-repositories (see also bottom of Fig. 4.5):

- The *Architecture Framework constructs* as described in Sect. 4.2: Obviously, in this Design Process, the Network Architect can use the Design Repository to retrieve the basic properties of these components as well as the transformation between them. E.g. after the Requirements Analysis phase, a first specification of the Strata is provided (the network architect is able to specify the main functions derived from the technical requirements that must be implemented in the horizontal strata, while the vertical strata should be always available as optional Strata that could be not present but that are recommended). According to the relationship between Strata and Netlets (also maintained in the Design Repository to be used by the Network Architect), once the Strata have been identified, the Network Architect can decide how to distribute the Netlets through the network nodes.

- The *Principles* that should be followed: e.g. how Strata can be composed, this will ease the specification of the interfaces and functionalities that maybe are not considered by just following the requirements. These principles are implicit in the specification of some building blocks but, thanks to the usage of specific modeling tools, the Network Architect can be supported in a more efficient way.
- *Design Patterns*: this repository represents the set of well known solutions for specific problems/requirements/functionalities that can be reused by the Network Architect. This might include, e.g., a specific protocol for the securitisation of an interface, a specific queueing algorithm for implementing a specific class of service, or a routing protocol (cf. Sect. 7.4.1).

4.9 Network Virtualisation Overview

As mentioned earlier, the concept of Network Virtualisation can constitute an enabling mechanism for actual deployment of new networks that were designed by the Network Architect within the Architectural Framework as described previously. Basically, Network Virtualisation is a concept to create logical network resources, i.e., *virtual nodes* and *virtual links*, to form a *virtual network* (*VNet*) from physical resources. The collection of physical resources is denoted as *substrate*, which is naturally divided into *substrate nodes* and *substrate links*. Some of the substrate nodes may offer virtualisation support and are therefore able to host virtual nodes, whereas some of them might not be able to host them. A substrate node with virtualisation support may host one or more *virtual nodes* of the same or different VNets. A *virtual link* is a link that connects two virtual nodes as shown in Fig. 4.6. A virtual link consists of a substrate path, i.e., a path that is composed of one or more substrate links. In general, a virtual link may consist of multiple substrate paths, which can be used to increase the capacity or reliability of the virtual link. Additionally, substrate path splitting [13] can be used to efficiently map VNets to substrate resources.

As depicted in Fig. 4.6, a VNet may span various substrate network domains belonging to different *Infrastructure Provider* (*InP*) networks. An immediate advantage for an operator of such a VNet is that it appears as a single homogeneous network even if it is actually composed of resources from heterogeneous InP networks. Finally, *end-users* will connect to the VNet infrastructure via virtual last mile links that also consist of substrate paths. In our view end user devices do not belong to the VNet topology as such, they are rather connected as leaves (or end nodes) to the VNet topology. Reasons for this decision will become clearer in the following, but the high dynamicity and the mobility of end users motivate their exclusion from the actual VNet topology description.

Like in the Internet today, one can assume that there will be multiple Infrastructure Providers (InPs) (Fig. 4.7), i.e., large companies that own the infrastructure required to enable communication between different locations and which provide end users with access to their networks. In contrast to today's Internet, the approach additionally considers availability of an inter-domain quality-of-service (QoS) solution in the substrate that can be used to establish links with QoS guarantees between

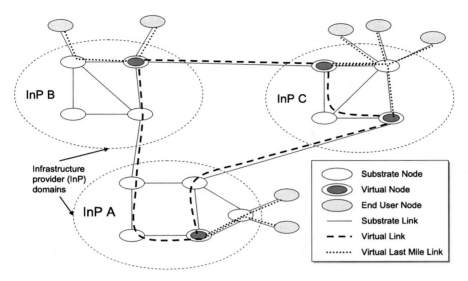

Fig. 4.6 Overview of a virtual network topology and substrate networks

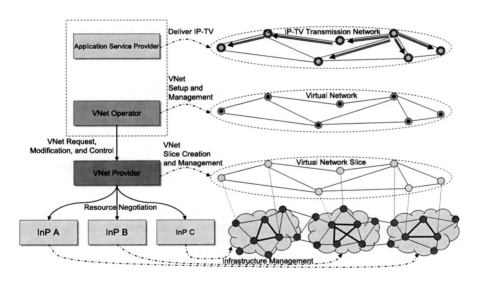

Fig. 4.7 Relationship between roles and resources

multiple InPs. Without QoS guarantees for virtual links, any further QoS mechanisms inside the virtual network itself cannot build upon a deterministic link capacity, and consequently QoS guarantees cannot be provided inside the virtual network.
InPs may also enable the creation of virtual nodes and virtual links on top of their
own resources and provide them to another party. In the VNet architecture, this other
party is the *VNet Provider*. The VNet Provider represents an intermediate party be-

tween VNet Operators and InPs and introduces a new level of indirection. A VNet
Provider composes a VNet *slice*, i.e., a lifeless set of virtual resources—as requested
by a VNet Operator—from physical resources of one or more InPs. This simplifies
the VNet operator's role, because he does not need to negotiate with different InPs.

After creation of the VNet topology, the *VNet Operator* can start to vivify his
network by installing and instantiating a network architecture inside and properly
configuring it. In some cases it is likely that an *application service provider* uses
the VNet to provide an application specific service, which in the chosen example is
an IP-based television (IP-TV) service. Usually, the IP-TV service provider would
specify the desired network topology for the VNet Operator, e.g., node locations
and capacities. The application service provider operates fully *inside* the VNet. One
or several *content providers* may access the VNet in order to provide the necessary
content, e.g., by putting TV programs and videos onto streaming servers that are
connected to the VNet. IP-TV service customers may then attach to the VNet and
use the provided IP-TV service wherever the VNet is accessible.

The InP provides the physical resources to construct the VNet, but he is not (and
does not need to be) aware of the content inside the virtual network. The VNet
Operator is responsible for the whole design of the service or application running
within the VNet, acting as a Network Architect: the Operator designs the network
architecture running inside the VNet by applying the Design Process. The same
network architecture might be running anywhere outside the virtual network as well.

Due to full end-to-end control over the VNet, the VNet Operator has the possibility to run protocols that support an application-specific service, e.g., in case of
IP-TV an IP multicast service may be necessary and needs to be deployed in the virtual network. In today's Internet the unclear business model and lack of deployment
of inter-domain multicast are severe obstacles for wide area IP-TV providers. Today, IP-TV is already in use, but usually offered only by InPs in their own domains.
An IP-TV application service provider could then use specialised intra-domain
multicast routing protocols, like PIM-SM (Protocol Independent Multicast—Sparse
Mode) and PIM-DM (PIM—Dense Mode), within his virtual network. The different
roles can also be combined, e.g., the VNet Provider and VNet Operator roles may
be represented by the same organisation, and in some cases the application service
provider role may be combined with the VNet Operator role. A fine-grained role
model, however, fosters a wide variety of business models that may not be realised
otherwise. This is an important aspect of the 4WARD VNet framework, because it
also covers usage in a commercial setting. But first the process of creating, setting
up and maintaining a VNet is described in more detail.

4.10 The VNet Lifecycle

In this section, an overview of the VNet lifecycle is presented, afterwards the identified interfaces in the VNet architecture are described. Such a "lifecycle" of a VNet
comprises the most important steps in the life of a VNet starting with the design up
to its operation. Normally, the destruction of the VNet would be considered in this

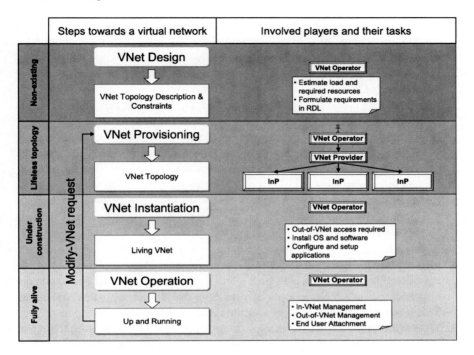

Fig. 4.8 The VNet lifecycle: Process overview

context, too, but this tear-down process is not in focus here. The VNet lifecycle is subdivided into the steps depicted in Fig. 4.8 as follows:

1. *VNet Design*: In order to create a new VNet, the VNet Operator has to describe the required topology, resources and corresponding additional constraints, for example, QoS constraints of virtual links or geographic restrictions of virtual nodes. The need to provide certain services or levels of service at specific points of presence is usually a cause for such geographic restrictions. In the particular example, the VNet Operator derives such specifications from the IP-TV application service provider requirements. It is therefore necessary to estimate the amount of virtual resources required to provide the intended service, but as the VNet can be shrunk or expanded later on, this only needs to be a coarse initial presetting and can be adapted at runtime. The VNet design integrates smoothly the Design Process and the resulting network architecture. During the creation, the actual requirement descriptions for the VNet will emerge as part of the final blueprint; and probably influenced by certain constraints imposed by the VNet Provider and/or the InP.

2. *VNet Provisioning*: The VNet description is then passed on to a VNet Provider who will construct the VNet from available physical resources at one or more InPs. The VNet Provider's main task is the construction of a VNet as described by the VNet Operator from Infrastructure Providers' resources by picking a set

of resources that matches the requirements. Therefore, he forwards the VNet description or parts of it to one or more InPs. An InP then tries to embed the requested VNet topology and resources into his substrate network, which is usually shared by different VNets. This embedding (or mapping) process consists of three steps [4]:

(a) *Candidate discovery and matching*: find a set of VNet candidates, i.e., appropriate virtual nodes and links, which fulfil the requirements.

(b) *Candidate selection*: chooses the best candidates using optimisation algorithms. InPs reply whether they can fulfil the request. The VNet Provider then decides which virtual resources should be set up and signals the binding in the next step.

(c) *Candidate binding*: allocates and reserves virtual resources from the substrate network to actually set up the selected candidates, i.e., each InP sets up the virtual resources.

A special case consists of interconnecting virtual nodes over InP boundaries, as the VNet Provider may have to assist in this task. Various cases have to be considered in this context, e.g., such a link may traverse InP domains that do not host nodes of this VNet or even involve InPs that do not support VNets at all. This is not a problem as long as a viable substrate path exists.

3. *VNet Instantiation*: If the VNet slice creation has been successful, the VNet Operator gets access to the virtual network slice. To allow the VNet Operator to enliven his share of resources he has to get access to an *Out-of-VNet Access* control interface. The functionalities offered by this interface must operate on a low level, e.g., allowing the VNet Operator to reboot the virtual machine in case of lockups, to install an operating system, and to access it similar to a serial console or remote control panel. Such an operating system could consist of the previously described Node Architecture. During creation of the VNet topology (i.e., a vivified VNet slice with a fully functional network architecture set up inside) or under severe failure conditions, management access is not possible from *within* the virtual network and therefore must be provided via an extra control plane interface, which we hence call Out-of-VNet Access.

4. *VNet Operation*: Some modifications of a VNet, e.g., extension, shrinking, modification of QoS requirements, or tear down of the VNet, may require contacting the VNet Provider again (cf. "Modify-VNet request" in Fig. 4.8). Other runtime operations without VNet Provider involvement include attachment of end users, virtual node migration (usually performed transparently at InP level) and controlled interaction with other VNets.

Figure 4.9 illustrates the following discussion on individual control interfaces. This figure shows the substrate consisting of three different InPs, i.e., InP A, InP B, and InP C. On top of their resources, a VNet has been created. The specific steps towards this virtual network are described in the following sections. In our IP-TV example, the Application Service Provider specifies the requirements of the content distribution network, e.g., node locations and link capacities.

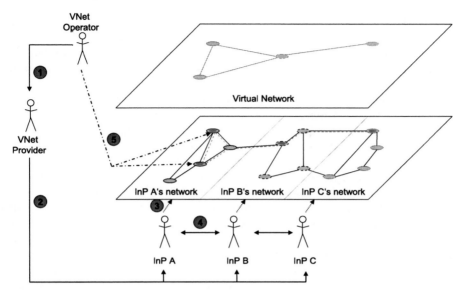

Fig. 4.9 The VNet lifecycle: Overview of interfaces

4.11 Creation of Virtual Networks

After the VNet Operator has finished the description of his virtual network (VNet Design Phase), the description is passed on to a VNet Provider (Interface 1 in Fig. 4.9). This interface may be assisted by proprietary Resource Description Languages (RDLs) and tool chains that VNet Providers might offer in order to compete with each other. For the interface between VNet Providers and InPs (Interface 2 in Fig. 4.9) however, this possibly proprietary description has to be mapped onto a common RDL that is used between VNet Providers and Infrastructure Providers. In order to keep this interface clean and to avoid that a VNet Provider has to translate the received virtual network description into multiple different RDLs, it is required to agree on a common, extensible RDL for this interface. More interesting than the actual RDL specification for this interface, however, is its signalling side. As this is a highly sensitive interface from a security and privacy aspect, it is isolated in a dedicated *Provisioning Network* that might be realised by a closed overlay network between VNet Providers and Infrastructure Providers. It is a closed network, because it should be accessible only by authorised VNet Providers and Infrastructure Providers. Furthermore, due to high availability and reliability requirements, it should provide high robustness against failures and attacks. Additionally, non-repudiation of transactions might be a desirable property.

Inside the Provisioning Network, a VNet Provider can then negotiate resources for virtual networks with multiple InPs. After the required virtual resources have been selected by the VNet Provider, each of the involved InPs starts to configure his own physical resources correspondingly and sets up all related resources for virtual nodes and links inside its domain (Interface 3 in Fig. 4.9). As this is an interface that is likely to be completely isolated from any other traffic in the Infrastructure Provider's domain, e.g., by using an isolated management network, InPs are free to use their own management tools. Nevertheless these management tools must be able to actualise all demands and constraints requested by the VNet Provider as agreed during resource negotiation.

Assuming multiple involved InPs, so far isolated parts of the virtual networks—each hosted at a different Infrastructure Provider—have to be interconnected (Interface 4 in Fig. 4.9). We note that the setup of virtual links between different Infrastructure Providers requires a common signalling protocol in the substrate and, in order to support quality-of-service (QoS) for virtual links, also calls for a common inter-domain QoS solution. In case of an IP-based substrate, a path-coupled resource reservation signalling protocol (e.g., QoS NSLP [7]) together with Diff-Serv mechanisms is used as a basis: resources are reserved along the substrate path that interconnects two virtual nodes. An extension of the QoS signalling protocol for negotiation and exchange of VNet specific addressing information is necessary, though. This Virtual Link Setup Protocol Object is an additional QoS NSLP object that is only interpreted by the end nodes of the substrate path. Integrating the resource reservation and signalling for virtual link setup in this manner reduces the setup time for virtual links. It is not possible to provide meaningful QoS guarantees on top of best-effort virtual links whose QoS parameters are not sufficiently predictable.

4.12 Instantiation and Management of Virtual Networks

After the virtual network topology has been successfully created, the VNet Operator can begin with the instantiation of the virtual network. As motivated before, this requires access to the virtual nodes, which has to take place *outside of the virtual network*. For instance, the VNet Operator may choose to use the same network operating system on all virtual nodes in order to simplify network management by using a homogeneous environment. Functionality of the Out-of-VNet access interface comprises low level management functions of the virtual node such as to start, to stop, to suspend, to reboot, or to install the virtual node and so on. This is an important control interface, especially if the instance running inside the virtual node cannot be contacted and managed anymore due to a failure inside the VNet, so that access from within the VNet ("In-VNet access") is not possible anymore. This *"Out-of-VNet access"* (Interface 5 in Fig. 4.9) may be provided either as direct access to the substrate node or as indirect access via a dedicated management proxy at the Infrastructure Provider. In general, one can assume that the VNet Provider provides a substrate contact address for Out-of-VNet management access to each virtual node, so both direct and indirect access are possible. In case of direct access and virtual node migration this information may be updated before the virtual node is actually moved to a new substrate node, whereas migration may happen transparently to the VNet Operator and VNet Provider behind a proxy for indirect Out-of-VNet access.

Depending on the role, there are different responsibilities during this phase of the VNet Lifecycle:

- *VNet Operator*—He performs *Out-of-VNet Management* to accomplish the above listed low level management functions and also *In-VNet Management* by using management mechanisms inside the virtual network. In case of an IP-based VNet the usual monitoring and management tools (e.g., SNMP, NetFlow, etc.) will probably be used. Monitoring of the virtual link QoS is also essential since the VNet Operator must indicate any violation of the service level agreements. For instance the number of transmitted and received packets over the virtual link can be monitored and compared in order to determine its current packet loss rate. Another responsibility is the handling of end user attachment, i.e., the VNet Operator must check a user's authorisation and delegate him to a suitable virtual access point.
- *VNet Provider*—He gets in the loop again if modification of the virtual network topology is required, e.g., due to adding or deleting virtual nodes. For instance, the IP-TV provider may want to increase the geographical coverage of his service and simply requests to add virtual nodes at new locations. Migration of virtual nodes between InPs also requires coordination by the virtual provider.
- *Infrastructure Provider*—He operates and manages substrate resources and optimises physical resource usage by taking advantage of virtual resource migration. Additionally, Infrastructure Providers offer the aforementioned Out-of-VNet access.

The interconnection of different VNets is possible by using so-called *Folding Points* as interconnection. A Folding Point can be composed of *Folding Links* and *Folding Nodes*. Different VNets may use different architectures, protocols, and addressing schemes, therefore requiring different translation mechanisms. In its simplest form, a Folding Link suffices for an interconnection of two VNets, e.g., when they are running the same architecture and protocols. The Folding Link is a special virtual link, because it is shared by both interconnected VNets. The Folding Node, however, may perform specific functions for the interconnection: e.g., security, authentication, policy enforcement, and translation of addresses, protocols and so on.

Interconnection of different VNets is also reflected within the Architecture Framework, as interoperability among heterogeneous networks, either virtual or physical, so the same principles apply. The interfaces defined among the different actors should find their proper implementation through the correct specification of the SGPs (Stratum Gateway Points) involved.

After a short description of a Virtualisation supporting substrate node architecture and touching some addressing issues in the next section, an interface for end-user attachment is briefly discussed.

4.13 Virtualisation Supporting Substrate Node Architecture

Every instance of a virtual node, hosted within the same or different physical node, could be structured as a whole Node Architecture, since it constitutes a complete op-

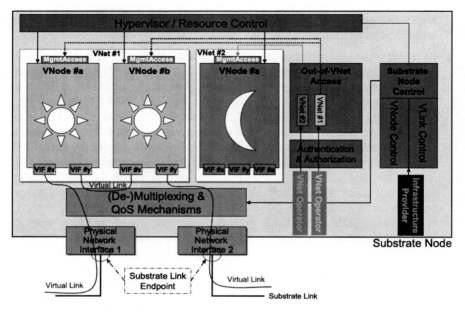

Fig. 4.10 A substrate node hosting different virtual nodes

erating node (cf. Fig. 12.5). The virtualisation supporting part of the substrate node architecture represents additional functionalities and interfaces, which are specifically sketched in Fig. 4.10.

While describing this figure, some requirements for identifiers required for signalling purposes are derived. Figure 4.10 shows a substrate node with two physical network interfaces. On top of this substrate node, virtual nodes of two different virtual networks are hosted: two nodes of VNet #1 (sun symbol) and one node of VNet #2 (moon symbol). For addressing purposes in control and data planes a unique *VNet-Identifier* (*VNet-ID*) is needed. It is required for multiple reasons:

1. *End User Attachment*: In order to attach to the desired virtual networks from any place, a globally unique identifier for virtual networks is useful, e.g., virtual access nodes of the corresponding VNet can be looked up and discovered (cf. Chap. 6).
2. *Accounting & Billing*: A globally unique identifier eases assignment of resource usage if multiple InPs are providing resources to a VNet.
3. *Uniqueness across multiple InPs*: Since VNets may span different InPs, the VNet-ID also should be globally unique, e.g., for accounting purposes. A VNet Provider generates the VNet-ID as a cryptographic ID, e.g., as hash value of a generated public key. This can be used for improving the security: in case a VNet Provider wants to modify a VNet configuration, InPs can verify that the VNet Provider possesses the corresponding private key that belongs to the aforementioned public key. Furthermore, the VNet Provider can supply credentials to the

VNet Operator and the involved Infrastructure Providers, so that only authorised access to control functions is possible.

If multiple virtual nodes of the same virtual network are hosted on the same substrate node—as is the case for VNet #1—it is necessary to differentiate between the virtual nodes, e.g., when setting up virtual links. Therefore, another identifier is required, the *VNode-ID*. The VNode-ID's scope is only valid with regard to a certain VNet-ID, i.e., it is possible that VNet #1 and VNet #2 consist of virtual nodes that are assigned the same VNode-ID (Fig. 4.10, VNode #a) as they can be differentiated by the VNet-ID. As a VNode may be connected via several virtual links to other virtual nodes, different virtual links may end inside a VNode that are accessible as a *virtual interface* (*VIf*) within a VNode. A virtual interface is thus identified by its *VIf-ID* that is unique inside the VNode. Figure 4.10 shows some important components:

- The *Substrate Node Control* consisting of VNode Control and VLink Control allows for setup and modification of virtual node slices and virtual links and must therefore only be accessible by the Infrastructure Provider.
- The *(De-)Multiplexing and QoS Mechanisms* component is responsible for de-multiplexing of multiple incoming/outgoing virtual links via one substrate link.
- The *Hypervisor/Resource Control* is responsible for actual creation of virtual nodes and manages the resources assigned to them.
- *Out-of-VNet Management Access* allows VNet Operators to access each of their virtual nodes in case of initial setup, misconfiguration, or failures inside the virtual network and permits reboot, serial console access, and further management functionalities. VNet Operators are allowed to access their virtual nodes after they have been properly authenticated and authorised. The access to this interface may be proxied by a management node of the Infrastructure Provider. From a security perspective, this interface is highly critical and requires extremely careful engineering.

Parts of the substrate node architecture were implemented based on XEN, Linux, and the Click Modular Router package. Evaluation results show that such substrate routers supporting virtualisation can be built based on commodity PC hardware while at the same time achieving a reasonable performance [3].

4.14 Virtual Links

With respect to the data plane, it is not required that the new identifiers are literally carried in data packets since there could be link-specific mapping techniques using available multiplexing mechanisms, e.g., 802.1Q VLAN-tags. In analogy *VNet Tags* denote such link-specific identifiers for VNets as an abstraction from concrete mechanisms. A VNet Tag identifies a virtual link of a VNet in a substrate link specific context, e.g., a virtual link of a certain virtual network might be mapped to Ethernet VLAN tag 42 in the substrate. This requires the presence of local mapping at the link ends. In absence of any substrate mechanism supporting (de-)multiplexing it may be required to carry an explicit shim header that carries a VNet Tag in order to allow for proper (de-)multiplexing of virtual links across a shared substrate link.

While virtualisation of servers, routers, wire line links, end nodes and hosts has been extensively studied in the literature, wireless link virtualisation has not yet received major consideration within today's research community. "Virtual Radio" is a framework that has been introduced by 4WARD dealing with the wireless virtualisation aspects. The framework focuses on configurable radio networks; further details can be found in [9].

Scheduling issues have been considered in the context of wireless virtualisation. Efficient scheduling of virtual links over shared last mile access networks has significant impacts on the quality of service for users and channel resource utilisation. In a system where a single wireless base station with a single antenna transmits delay tolerant data to multiple virtual networks, when the partial Channel State Information is available, the optimal scheduling strategy to maximise spectrum efficiency is to transmit to a single user with the best channel quality in each scheduling epoch (i.e., time slot) [6, 10]. Furthermore, a performance analysis for a variable number of VNOs (with several types of traffic) accessing the same wireless medium through the same wireless interface (on a Time Division Multiple Access—TDMA—basis) was conducted. The core algorithm of the WMVF (Wireless Medium Virtualisation Framework) is based on the Weighted Round Robin scheduling mechanism, modified with some specific rules and enhanced to become an adaptive technique. Adaptation enhances the performance of the main principle (TDMA), since the access node can measure (in principle) the usage conditions taken by the different VNOs and adapt the assignment of their Time Slots in the WRR accordingly. The idea is to look for the maximum usage maintaining the best possible QoS conditions for each traffic type. The Cooperative VNet Radio Resource Management (CVRRM) is devoted to manage the radio resources among VNets, in a heterogeneous environment. Its main objective is the analysis of a set of different virtualised wireless technologies, instead of looking into the virtualisation of a particular technology.

4.15 End User Attachment to Virtual Networks

If an end-user (e.g., an IP-TV customer) attaches to a substrate network, he probably wants to access the dedicated IP-TV VNet (or even several VNets) to which he subscribed. Rather than having to explicitly "dial-in" into the virtual network by establishing a tunnel connection to a VNet concentrator (analogous to a VPN concentrator), connectivity should be established automatically, if possible. That is, the end-user's node will automatically discover virtual access points of the VNets it wants to connect to. A *Virtual Network Attachment Protocol* is used to contact the VNet Operator of the corresponding VNet for initial authentication and authorisation. The end-user is probably attached to a domain that either does not support Network Virtualisation or does not have any virtual nodes belonging to the specific VNet. Thus, the substrate access node needs to support only some generic backend authorisation protocol. Two alternatives exist in this context: Either the substrate access node locates the corresponding authorisation node of the VNet Operator for the requested VNet or the request carries this information already. The response contains the substrate address of the VNet Access Node that can be used for the subsequent setup of the point-to-point virtual last mile link. In a simple case one could think of tunnel creation, in more complex cases the virtual last mile link may offer guarantees and a QoS reservation is required. The VNet end-user could thus access his subscribed IP-TV service from nearly any location. In case that VNet technology will be heavily used in the future, an end-user would automatically connect to several different VNets for access to the various subscribed services or network architectures. Current network access technologies need to be extended to provide such a function for the expected multitude of virtual networks.

4.16 Conclusions

In the context of the Future Internet it is envisioned that different network architectures can co-exist and share a common infrastructure. These network architectures can be specifically tailored to particular user or application requirements and, furthermore, can take into account the characteristics of the available networking resources.

Therefore, the design of new network architectures is simplified and the productivity of those designing such architectures (called network architects) is expected to increase considerably. Overall this may have a high impact on new innovations on the socio-economic side. Developing network architectures may no longer be a long and painful act of slow-moving standardisation bodies. Moreover, the provisioning of architectural principles, design patterns and building blocks may lead to the deployment of new business cases. VNets provide a versatile platform to ease such a deployment.

In this chapter a proposal has been presented which describes a possible Architecture Framework covering certain demands of the Future Internet. The specification of the concepts being developed to form this framework is elaborated in much more detail in [1].

To model new Network Architectures, some concepts, terms and the basic constructs have been defined. This Architecture Framework provides two levels of views on network architectures: (i) the macroscopic view mainly focuses on structuring the network at a higher level of abstraction and introduces the concept of Strata as a flexible way to layer the services of the network that can enable the usage of information across different layers; and (ii) the microscopic view concentrates more on the functions needed in the network nodes, their selection and composition to Netlets that are instantiated in the Node Architecture that allows the dynamic coupling of applications and network protocols not easily possible in today's networks. The Functional Blocks are presented as the common points between the two views of the architecture. The Component Based Architecture constructs and principles are used as the basis to provide reusable frameworks that minimise the design and development times of new network architectures.

The methodology to be followed is represented in the Design Process; which considers 3 main phases: (i) detailed requirement analysis, (ii) the Abstract Service Design and (iii) the Component Design Phase. This proposal does not aim to substitute the current design processes available in different organisations but it presents a proposal which links the communication system design and the software development principles. The Design Process is complemented with a Design Repository which provides guidelines, design patterns and the like, in order to support the Network Architect.

Virtual networks open up a migration path to new network architectures, which can be built according to the previously described Design Process, and also introduce new flexibility for InPs with respect to their resource management. The presented VNet framework considers the use of virtual networks in a commercial setting. In contrast to other approaches the architecture considers four involved players: the Infrastructure Provider, the VNet Provider, the VNet Operator, and finally the VNet end-user. The framework covers the different necessary signalling and management interfaces that are required during the presented lifecycle of a VNet. Although there are already many virtualisation techniques and mechanisms in wide use today, there is still some work needed in order to realise the sketched fully featured VNet architectural framework. Especially, coordinated interaction between different InPs (e.g., for setting up virtual links) requires standardisation of the respective signalling interfaces. Parts of the presented VNet architecture were realised and evaluated in form of individual feasibility tests.

References

1. M.A. Callejo, M. Zitterbart et al., 4WARD D2.2 Draft Architectural Framework (2009), http://www.4ward-project.eu/index.php?s=file/s/do5(d)ownload&id=35, Document Number: FP7-ICT-2007-1-216041-4WARD/D-2.2
2. I. Crnkovic, J.A. Stafford, H.W. Schmidt, K. Wallnau (eds.), *Component-Based Software Engineering, 7th International Symposium, CBSE*, Edinburgh, UK, May 2004
3. N. Egi, A. Greenhalgh, M. Handley, M. Hoerdt, F. Huici, L. Mathy, Towards high performance virtual routers on commodity hardware, in *CoNEXT'08: Proceedings of the 2008 ACM CoNEXT Conference* (ACM, New York, 2008), pp. 1–12, http://doi.acm.org/10.1145/1544012.1544032
4. N. Feamster, L. Gao, J. Rexford, How to lease the internet in your spare time, SIGCOMM Comput. Commun. Rev. **37**(1), 61–64 (2007), http://doi.acm.org/10.1145/1198255.1198265
5. FP7 4WARD Project (2009), http://www.4ward-project.eu/
6. I. Houidi, W. Louati, D. Zeghlache, A distributed virtual network mapping algorithm, in *IEEE International Conference on Communications (ICC'08)* (2008), pp. 5634–5640, doi:10.1109/ICC.2008.1056
7. J. Manner, G. Karagiannis, A. McDonald, NSIS Signaling Layer Protocol (NSLP) for quality-of-service signaling, RFC 5974 (Experimental) (2010), http://www.ietf.org/rfc/rfc5974.txt
8. D. Martin, H. Backhaus, L. Völker, H. Wippel, P. Baumung, B. Behringer, M. Zitterbart, Designing and running concurrent future networks (demo), in *34th IEEE Conference on Local Computer Networks (LCN 2009)*, Zurich, Switzerland, Oct. 2009

9. J. Sachs, S. Baucke, Virtual radio: A framework for configurable radio networks, in *International Wireless Internet Conference* (*WICON 2008*), Maui, USA, 2008

10. D.N.C. Tse, Optimal power allocation over parallel Gaussian channels, in *Proc. of Int. Symp. Inf. Theory* (1997), p. 27

11. L. Völker, D. Martin, I. El Khayat, C. Werle, M. Zitterbart, A node architecture for 1000 future networks, in *Proceedings of the International Workshop on the Network of the Future 2009* (IEEE, Dresden, June 2009)

12. L. Völker, D. Martin, C. Werle, M. Zitterbart, I. El Khayat, Selecting concurrent network architectures at runtime, in *Proceedings of the IEEE International Conference on Communications* (*ICC 2009*) (IEEE Communication Society, Dresden, June 2009)

13. M. Yu, Y. Yi, J. Rexford, M. Chiang, Rethinking virtual network embedding: Substrate support for path splitting and migration, SIGCOMM Comput. Commun. Rev. **38**(2), 17–29 (2008), http://doi.acm.org/10.1145/1355734.1355737

Chapter 5
Naming and Addressing

Some Remarks on a Basic Networking Ingredient

**Holger Karl, Thorsten Biermann,
and Hagen Woesner**

Abstract A discussion of "names", "addresses", and their relation to each other from first principles is provided. We put this discussion into the context of the structure of communication systems, where we compare layered schemes vs. functionally complete ones. We identify the need for namespaces with proper information hiding and how to relate such namespaces to each other by a correctly interpreted name resolution concept. It turns out that this approach simplifies a number of aspects of communication system design; for example, it turns out that name resolution and neighbor discovery are essentially the same thing. A basic view is given on names, addresses, and compartments. Name resolution is presented as the centerpiece of the problem.

5.1 The Role of Names and Addresses

Few concepts so easily create confusion as the seemingly innocent terms "names" and "addresses". Yet few other concepts are so basic to networking as these two, along with their relation to each other, and few others have so many different notions and semantics attached to them in different communication systems.

But is this diversity just sloppiness on the side of the networking community, or is this the reflection of a deeper need to identify entities at different levels of

H. Karl (✉) · T. Biermann
University of Paderborn, Paderborn, Germany
e-mail: hkarl@ieee.org

T. Biermann
e-mail: thorsten.biermann@upb.de

H. Woesner
Technical University of Berlin & EICT, Berlin, Germany
e-mail: hagen.woesner@eict.de

L.M. Correia et al. (eds.), *Architecture and Design for the Future Internet*,
Signals and Communication Technology,
DOI 10.1007/978-90-481-9346-2_5, © Springer Science+Business Media B.V. 2011

abstractions? Is it plausible that this diversity can be overcome by a more stringent attempt at defining these term, or is it a natural consequence of the problem space; a consequence any networking architecture has to handle? This chapter attempts to provide a possible perspective on this question, as developed by parts of the 4WARD project.

5.2 A Basic View: Names, Addresses, and Compartments

5.2.1 What Is a Name, an Address?

Definitions for names and addresses abound in the literature. Let us consider a few examples.

> In a distributed system, names are used to refer to a wide variety of resources such as computers, services, remote objects and files, as well as to users. [1, p. 368]

This observation immediately points out that names take on many different shapes, by the simple fact that they are applied to vastly different entities. A networking architecture that imposed a single form of names would be highly counter-intuitive.

Shoch gave a succinct characterization of the relationship between names and addresses and also includes the notion of a route:

> The 'name' of a resource indicates *what* we seek, an 'address' indicates *where* it is, and a 'route' tells us *how to get there*. [7]

He continues by defining a name as the identifier of a (set of) resource(s); however, he considers a name to be something typically human-readable. Then, the discussion of "addresses" becomes less clear: An address is considered to define the "fundamental addressable object"; it must also be understood by all participants (at least, its format). Both these properties can be justifiably be regarded as a property of a name as well, making the distinction between these two concepts less clear than what was contained in the brief characterization above. Shoch also considers two mapping processes, one that maps a name to one or several addresses (possibly with implicit notions of a preferred address out of a set), and one mapping from an address to the route. Both these mappings can change over time and need to be known a priori, but can be determined only at communication time. While not explicitly stated, Shoch seems to imply that these two mappings are separate steps (a route is defined as the information needed to forward information to its *specified* address). We shall later see (Sect. 5.3.5) that it can make a lot of sense to determine both the name/address mapping and the address/route mapping simultaneously. Judging from his examples (i.e., the telephone system and computer networks), Shoch seems to regard name and address as strictly different things, but he does consider the issue of where to place this boundary in the example of a distributed computing system.

The discussion by Saltzer [6] goes a step further by distinguishing more finely between four different kinds of objects that can be named: services/users, nodes, network attachment point, and paths. Between these different objects, name mappings are assumed. He observes that just trying to distinguish between names and addresses will not lead to much insight; a lesson learned from the operating system community. Rather, any kind of identifier (be it a name, an address, or other) is usually bound to another identifier *within a given context*. Saltzer points out that the form one chooses to represent an identifier has no bearing on its semantics: just because an identifier is printable or human-readable does not make it a name; just because it is a binary format does not make it an address.

The key point of Saltzer (for the point of our discussion) is that names appear at various abstraction levels. While he sticks to concrete examples, it seems promising (and obvious) to generalize this notion beyond the four concrete classes of objects identified by Saltzer. Nonetheless, the intuition of "an address describes the *where* of an object" needs to be captured as well if any terminology should make sense.

A more recent example of a basic naming/addressing discussion can be found in Day's book [2]. He follows along the lines of Shoch and Saltzer, yet sticks to rather conservative assignment of names to applications and addresses to "lower layer" entities [2, p. 158]. We consider such a rigid assignment to be disadvantageous.

5.2.2 Some Structural Aspects of Name

Names, in the abstract, are identifiers for entities. Within a given context (later on defined as a "compartment"), concrete names come from a set of possible names. Such a set is called a *namespace*; it stipulates certain rules and structural aspects how its names can be formed and what operations on names are allowed. As an example, a namespace can define the notion of *identity* between names (a property which we will require of all namespaces considered here).

Some aspects of names and namespaces repeatedly come up in discussions about names and often confuse such discussions rather than to add insight. Some of these aspects are the following:

Hierarchical or flat A namespace can use "flat" names that do not carry any discernible structure. Or the names could follow some hierarchical arrangement, possibly only to some degree or in parts of the names.
Hierarchical names lend themselves easily to the construction of aggregated routing schemes, but this is not a necessary condition. Approaches to aggregated routing even with flat names exist; Bloom filters are typical examples.

Opaque or transparent Are names opaque or transparent, i.e., do they hide or reveal their inner structure (if any)? Are members of a namespace treated differently than non-members in this respect (typically, there is a difference as namespace members can understand name structure).

With or without name allocation Are names allocated only by the entity itself, or is some consensus process with other entities necessary? Is this process distributed or centralized?

With or without admission control Is obtaining a name in some way controlled? Are preliminary names assigned during this procedure?

Explicit or implicit Are names explicitly used or are implicit names allowed, e.g., functions that evaluate to a (set of) entities during usage. If the latter, where and when does such a function evaluation take place?

Unique or not Are names required to be unique, i.e., can a name be assigned to more than one entity? If uniqueness is required, how is it enforced, how does this relate to name allocation? If uniqueness is not required, what is the semantics of having the same name assigned to multiple entities (an anycast, multicast, ... semantics)?

Persistent or not Are names persistent, i.e., can the name of an entity change?

Reusable or not Are names reusable, e.g., after an entity assigned a given name no longer exists?

Relinquishable or not Can an entity release a name?

Wildcards Do wildcards exist, i.e., names that apply to an anonymous or implicitly defined set of entities? Are there more than one wildcard, and if so, what is their semantic difference?

Anonymous entities allowed Is an entity allowed to not carry a name? Is this only allowed as a transient state while an entity obtains a name? Does such an entity still answer to certain wildcard names?

Individual or group names Do group names exist, or do names only apply to a single entity? What is the semantics of a group name, are all, one, or some members of a group designated by it? How does this relate to uniqueness of names (e.g., there might be unique names per entities and unique names per group, yet still entities might belong to several groups)?

More than one name per entity Is an entity allowed to have more than one name (at a time, or at all)?

Security aspects Do names carry security properties, e.g., are they self-certifying?

We are striving here for a concept that can encompass all such aspects; we will try to make as few limitations on the structure and semantics of a set of names as possible. It will turn out during this chapter that we insist on names being opaque outside their namespace—i.e., an entity that is not member of a namespace is not able to understand meaning and structure of such names. Also, a wildcard name is highly useful but not mandatory. Other than that, we believe we do not need to make any further assumptions on the structure of names.

5.2.3 Structures in Communication Systems

No existing communication system contains all required functions in a single implementation block. All systems use some form of internal structuring to hold the conceptional and implementation complexity at bay.

5.2.3.1 Layers

Typically, these structures are *layers*: Conceptional units of restricted functionality, each layer provides a specific, abstract view of a communication system. Usually, layers are arranged in increasing order of convenience of this abstract system view. A layered system is usually distinguished from an otherwise structured communication system by (a) specific functions being realized in only a single layer (a few functions, like error and flow control, might appear in two functions), (b) strict information hiding between these layers, and (c) a rule that only immediately adjacent layers are allowed to interact.

Cross-layer optimization (see, e.g., [8] on wireless cross-layer optimization) relaxes the opaqueness of layers and allows interactions also between non-adjacent layers. While that approach might harbor great benefits and has sparked lots of research, it is largely immaterial for the discussion at hand.

A common misconception of the layered architecture as such and of its typical example, the ISO/OSI 7-layer stack, is the following: Since routing and forwarding only happen at a single layer (even that is only true for the purest possible form of the 7-layer stack, but usually not true in practice), only at a single layer is there a need to think about names and, in particular, addresses. While it is possible to take such a stance in principle, we shall see below that a broader view of routing, forwarding, names, and addresses can both enrich and simplify networking architectures.

5.2.3.2 Functionally Complete Structures—DIF

Going beyond just cross-layer optimization, a more fundamental change is to allow each structural unit to be *functionally complete*, i.e., each structure is allowed to implement all possible functions of a communication network. At first glance, this does not ease the conceptional or implementation challenge of a communication system. The key observation is here that it still holds that these structures are implemented on top of other such functionally complete structures, with more or less convenient properties. In addition, a key difference is also that these various structural levels operate at different *scopes* of a real system. Obviously, it then becomes possible to apply this idea recursively; at each recursive step, the scope of operation, the applied policies, the chosen protocols can be chosen independently.

One example of such a structure is the Distributed IPC Facility (DIF) concept according to Day's NIPCA architecture [2]. The emphasis is on pointing out the similarity of network communication with interprocess communication. The terminology is that of a system with a strictly ordered sequence of DIFs allowed to call each other (akin to a layered system); there does not, however, appear to be any reason not to allow "cross-DIF" optimization. For a DIF, there is a fairly complex set of concepts regarding applications, application protocol machines, instances of these protocol machines, ports, and the distributed application in general; each of these concepts has a name or, rather, an Identifier (ID) assigned to it. A crucial observation is that a DIF consists of a small set of basic functions; in particular, the *relaying and multiplexing task*, the *error and flow control protocol*, a *resource information exchange protocol*, and an *access protocol*. The relaying and multiplexing task and the access protocol are closely linked to the naming discussion, as access protocol assigns names and the multiplexer needs to identify the correct peer entity to which to pass on a data item.

A main point to conclude is that in a DIF-based architecture, names appear at each DIF, and so does routing and forwarding. This is in sharp contrast to functional layering.

5.2.3.3 Compartments

To a large extent we agree with the basic notions and concepts of the DIF approach. We emphasize the need to distinguish DIFs that are very similar, but can have boundaries between them for technical, administrative, legal, business, or other reasons. We borrow the term and concept *compartment* for such a DIF from the Autonomic Network Architecture (ANA) project [3] and define it, for our purposes, as follows (see [5] for details):

- A compartment is a (possibly empty) set of entities, all able to use the same communication protocols.
  In particular, it is acceptable for a compartment to include more than one communication protocol, as long as any entity belonging to the compartment understands all these protocols.
- A compartment has an associated *namespace* containing possible names for these entities.
  In particular, we do not require entities to have unique names; we also allow an "empty" name to be included in the namespace (representing the case when an entity does not have an actual name). Moreover, a namespace may carry operations on names, e.g., "included in", "imply by", or others. The test for equality between two names must be defined by a namespace.
- We also require that all entities inside a compartment are *in principle* able to communicate with each other; details of this ability are described by a set of rules and requirements.
  This pertains, in particular, to the *technical ability* to communicate. What constitutes an acceptable level of communication ability is detailed in the set of requirements; the typically intended notion is that short-term error events do not jeopardize compartment membership whereas long-term changes of communication quality do. As an example, consider a mobile ad hoc network, starting out at as a single compartment. When nodes move out of communication range, the compartment fragments into two (or more) compartments. They might later be recombined into one when nodes come again into mutual reachability.
  This rule set also pertains to *administrative rules*, defining whether a direct interaction between two entities is even desired. For example, two entities might be technically able to communicate, have names from the same namespace, and understand the same protocols—yet are not allowed to communicate directly because they belong to different business entities or administrative domains (for

example, WLAN devices might well be in mutual radio reach but are not allowed to talk to each other; different SSIDs or encryption are examples for technical means to ensure that).

Note that addresses are not mentioned in this definition.

A consequence of this notion is that entities belong to different compartments are in fact not able to communicate. Because, if they are in different compartments, one of the three requirements above would not be met, hence, they cannot communicate. At first glance, this might seem odd, but it in fact captures immediately the idea of ability to exchange semantically meaningful information. The reader should also not confuse this with gatewaying: at some point, there is an (maybe even only implicit) compartment to which the actually communicating entities belong.

5.2.3.4 Generic Paths

The notion of an *entity*, details about rule sets, and the communication protocols are further detailed in Ref. [5]. This reference describes the Generic Path architecture provides an abstraction for data transport across and/or data manipulation inside a network facility. It provides object-oriented means to defines classes of such paths and to instantiate such paths. It is also able to assign identifiers to routes in the sense of Shoch [7] and Saltzer [6] if so desired. Moreover, this architecture distinguishes between the *endpoint* of such a path and an *entity*, which groups multiple such endpoints together and is the unit of participation in a compartment (it is perhaps easiest to think about endpoints as the finite state machine for a concrete data flow and the entity as the collection of these flows along with the necessary logic and state to set up, control, and tear down such endpoints).

For brevity, we restrict the discussion to crucial points and refer the interested reader to Ref. [5]. In particular, the remainder of the discussion in this chapter will gloss over (important) details in the difference between entities and endpoints and how they relate to each other with respect to the setup of data flows.

5.2.4 Names and Addresses vs. Structures

5.2.4.1 Addresses Come from the Outside

What does this discussion tell us about the relationship of names on the one hand, addresses on the other hand, and how these two terms relate to structures in communication systems? The important observation has essentially been made clear by Saltzer and Shoch already, and it is clear when looking at any textbook on communication systems:

> Inside a compartment, there are no addresses, only names.

The textbook argument is an obvious one. When defining the routing problem on a simple graph, there is only one kind of identifier, a name for a node. There is no need for an "address" in the context of routing on such a graph. When looking at a multi-graph (where two nodes may be connected by more than a single edge), there is a need to distinguish between these different options to reach a neighboring node in this graph, and this need usually arises because these different edges may have different costs. Yet the routing still happens on the basis of names along the nodes of the graph.

From where, then, do addresses come from? An example of everyday usage quickly clarifies the confusion. The *name* of a person might be "John Doe". An *address* of this person might be "Main Street 1". But this address also comes from a namespace, the one that is used to designate buildings. Focusing on this namespace, it is clear that "Main Street 1" is actually a *name*, the name of a building. Another address of John Doe might in fact be "+01 123 456 7891"—the *name* of a telephone. Hence, John Doe might have many addresses, each of them are names of entities in different compartments.

This observation immediately carries over to a communication system and allows us to give the following definition:

> An *address* of an entity E_1 in compartment C_1 is the name of another entity E_2 in (typically) another compartment C_2. The name of E_2 is turned into an address of E_1 by *binding* entity E_1 to the desired name of E_2.

Casting this in the light of Shoch's definition of name vs. addresses (Sect. 5.2.1), we see that an address indeed tells us "where" a name is, but tells us so *with respect to another namespace*. Looking only at a single compartment, there is no need nor does it make sense to talk about an address; inside a compartment, names and their neighborhood relationship are fully sufficient to define a topology (a graph is defined as a set of node names and edges between these names; there is no third concept of an "address" necessary).

This relationship is illustrated in Fig. 5.1 where the entity E_2 has two names $N_{2.1}$ and $N_{2.2}$. Entity E_1 is bound to name $N_{2.1}$, turning this name into an address for E_1. If confusion might arise, we will sometimes include the compartment from where the address originated: $N_{2.1}@C_2$ is the address of E_1.

From the perspective of the compartment C_2, $N_{2.1}$ is still a name. In fact, from the perspective of C_2 or E_2, little has changed.

Just like an entity might have several names inside its compartment, it might have several bindings to addresses, stemming from different compartments or from different entities inside the same compartment. As an example, let us look at an entity inside an "IP" compartment (say, an IP routing engine). It carries one or several names (its so-called "IP addresses", a serious misnomer in the presented terminology). This entity is also bound to the name of several entities in an "Ethernet" compartment as well as to an entity in a WLAN compartment.

Fig. 5.1 The entity E_1 in compartment C_1 is bound to the name $N_{2.1}$ of E_2 inside C_2, making $N_{2.1}@C_2$ an address of E_1

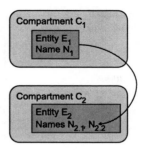

In summary, whether a given identifier is a name or an address is a matter of perspective. Inside the compartment (or namespace) to which this identifier belongs, it is *always* a name. Once an identifier is bound to an entity outside of its compartment, it can be regarded as the address of the *foreign* entity. A name is *never* an address inside its own namespace; such an expression is senseless.[1]

5.2.4.2 Basic Communication: The Need for a "Node Compartment"

The purpose of an address binding is, eventually, to have data flowing between the entity which provides its name as an address (in Fig. 5.1, E_2) and the entity which binds itself to this address (in Fig. 5.1, E_1). In Sect. 5.2.3.3, we learned that *only* entities that share some compartment can communicate with each other.

This seems like a catch 22: Entities in different compartments cannot communicate, but to realize communication, we need different compartments (in the sense of simpler to more complex functionality or smaller to wider scope). We can break this cycle by observing that entities naturally share an environment that fulfills are the requirements of a compartment, namely, a typical operating system.

- Two entities existing inside the same operating system share some form of communication protocol, namely, the Interprocess Communication (IPC) functions of the operating system.
- An operating system provides a particular namespace (e.g., process identifiers, thread identifiers, or pointers to finite state machine automata) which can be used to name an entity inside it.
- Via the operating system IPC, two entities are in principle able to communicate; whether they are allowed to or not (e.g., have sufficient rights) is a decision to be taken by the compartment, viz., the operating system.

---

[1] We note in passing that we deviate here considerably from the NIPCA architecture. Day writes, e.g.,

> The addresses must be large enough to name all elements that can be communicated without relaying at the layer above. [2]

Hence, he regards names and addresses as fairly interchangeable concepts. In the terminology presented here, a phrase like "an address names something" would not be considered correct.

Fig. 5.2 The entities E_1 and E_2 share a single node compartment

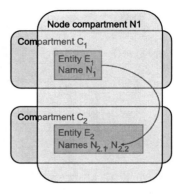

With this observation, we need to extend Fig. 5.1 by introducing the node compartment, enclosing the two entities E_1 and E_2 (Fig. 5.2). We also conclude the following crucial observation:

> Bindings of names to addresses are only possible between entities sharing a node compartment.[2]

For practical purposes, it is often necessary to find out whether a binding exists between an entity and another entity (of the same node compartment), or to find all possible bindings of a given entity. This need arises, for example, in the crucial step of name resolution, when possible ways to find a given name are considered. Another example is mobility of an entity (e.g., mobility between node compartments), when it becomes necessary to find out whether any bindings would be destroyed by moving an entity around (and which, if any, measures have to be taken). In principle, this information is available by inquiring all involved entities, but this can become an considerable run-time overhead. It is hence a reasonable *implementation* choice to collect all this information in a *binding table*, which collects all this information inside a node compartment (conceptually, this information is not necessary as it can always be reconstructed by asking all entities).

5.2.4.3 Information Hiding and Name/Address Binding

Insisting, as we do, on a strict separation between names and addresses has a number of consequences, mostly related to information hiding.

The first and most important consequence derives from our postulate that names carry semantics which can only be understood and processed *inside* their own

---

[2]In principle, *any* shared compartment (not necessarily a node compartment) enables address binding. The practical cases for such bindings turn out to be rather esoteric, so we ignore them for the present discussion.

namespace, more precisely, by entities belong to a compartment using that namespace. But when a name of one namespace is used as an address for an entity of another namespace, the so-addressed entity is unable to understand the semantics of its own address. An entity can therefore, in general, not process its own address(es); they are meaningless, opaque bitstrings for an entity.[3]

While this might look like a problem, it is in fact a considerable advantage. It prevents modifying data (= addresses) outside of an entities purview and makes the need to bind entities to addresses explicit. This binding can be reflected at carefully chosen locations (e.g., once per node; see [5] for details), giving a natural place to control, for example, mobility. Moreover, with a proper software engineering approach, it is even possible to prevent addresses from appearing inside the payload of a given compartment (where they have no understandable semantics anyway), circumventing some of the well-known problems of current protocols (e.g., IP addresses being used inside application protocol payloads like in FTP or SIP).

A second consequence is the need and the possibility to control the naming and binding process. This pertains to both (a) how names are obtained by an entity, (b) to which compartments and entities an entity might obtain a binding, (c) how are bindings made known, and (d) how are existing bindings found.

The first point—how names are obtained—is strictly a compartment-internal decision, governed by the compartment's own rules. For example, a compartment can proscribe that names must be unique and that an entity has to obtain authorization to use a name; it could proscribe a centralized assignment of names; a compartment could also be entirely permissive and not impose any rules on how names are used (requiring proscriptions for name collisions). These rules are captured by a compartment's *name allocation protocol*. The name allocation as such does not yet imply any bindings; creating a binding is a separate, intentional step by the entities. There are good reasons to separate these two steps; just consider the case where a name is allocated using one assisting compartment but the actual binding shall be done to entities in another assisting compartment.

The second point—how bindings are obtained—is a more complex affair. Creating a binding would typically be initiated by an entity looking for an address. Obtaining an address means that an entity (can) become reachable via the compartment from which it has obtained the address. Such a decision can depend on *admission control procedures* imposed by either of the two compartments; it could also depend on the execution context where these two entities exist (typically, the operating system). For example, while both a voice application entity and an WLAN entity might be willing to set up an address binding between each other, the operating system of a mobile phone might prevent this binding, in order to force the voice application to create a binding with an entity from a cellular networking compartment.

The third point—how are bindings made known—and the fourth point—how to find out an existing binding—are strongly interrelated. Obviously, there is a wide range of possible approaches, ranging from fully centralized to fully distributed with

---

[3] We ignore here the degenerate case of an entity obtaining an address from another entity inside its own namespace.

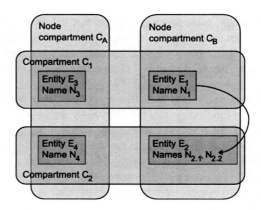

Fig. 5.3 Address serve to find a way to a neighbor

many hybrid solutions in between. Section 5.3 will look at some example realizations, but the important point here is *where* with respect to the compartment structure this information should be stored and *who* needs to access it.

5.2.4.4 Neighborhood

Existing name/address bindings have to be consulted to determine via which compartment a given entity can be reached. We did assume earlier that inside a compartment, mutual reachability is a defining characteristic. However, in reality, two entities in an arbitrary compartment cannot directly communicate using primitive means, but rather have to rely on other, simpler compartments to convey information from one entity to another; this is one aspect of layering models of communication systems. Hence, an entity A needs to know an address of entity B (both belonging to the same compartment) in order to determine whether B can be considered a *neighbor* of A. To do so, A needs to know the name of B and both A and B need to have access to the same, *assisting* compartment from which the address of B originates.

Figure 5.3 illustrates this point. Like before, entity E_1 has bound itself to the address $N_{2.2}@C_2$. This address can be used to direct a message toward E_1, using the help of entities in the compartment C_2 to actually transport this message. Obviously, which kind of compartment C_2 is, as such, not relevant if only it can communicate messages between two entities of C_1. If so desired, it is possible to require some kind of semantics or quality from the communication facility, represented by C_2. Such requirements can be expressed in various form; we again refer the interested reader to Ref. [5] for details.

The important point to note here is that by means of bindings between entities in *different* compartments, the notion of *neighborhood* in a given compartment can be defined. From common intuition, two entities inside a given compartment are neighbors if they can communicate directly, from the perspective of that compartment, without the help of any other entities of this compartment. This could be the case for some simple compartments even without the help of any further communication service; the typical examples here are the node compartment or compartments close to physical transmission. For other compartments, neighborhood requires the help of assisting compartments. Hence, we can define neighborhood of two entities quite formally:

Two entities E_1 and E_2 both belonging to some compartment C_1 are called

- C_1 *neighbors* if they can communicate directly with each other if they can do so without the help of any other entity or any other compartment;
- C_1 *neighbors with respect to* C_2 if there exists one additional compartment C_2 with entities E'_1 and E'_2 such that E_1 and E'_1 can communicate and E_2 and E'_2 can communicate with each other and E'_1 and E'_2 are willing to carry traffic on behalf of E_1 and E_2 (they are said to "offer a communication service").

At first glance, this definition looks like an unfounded recursion: how could E_1 and E'_1 (and, similarly, E_2 and E'_2) in turn communicate; this is even more difficult than the communication between E_1 and E_2 in the first place? Indeed, this is only possible under a certain assumption: both E_1 and E'_1 have to share one compartment. And there is a natural compartment which can fulfill this role: the node compartment. In this compartment, we assumed the ability to communicate to be given as a basis for all further communication systems to build upon (in a sense, similar to Day's motto of "all communication is IPC"). This does *not* preclude, however, to build more complex arrangements of compartments via which service offerings can be made; we simply restrict the further discussion to this case. Similarly, there are some compartments (namely, the "physical layer" ones) where communication between some entities is naturally defined, by physical phenomena. These also serve as fundamental cases for the remainder of the discussion.

5.2.4.5 Neighbor Discovery

Neighbor Discovery in an Arbitrary Compartment We have now defined the notion of neighbors inside a compartment. Closely related is the notion of *neighbor discovery*. For an entity E_1 in a compartment C_1 to find out about its possible neighbors, this is assumed given for C_1 neighbors directly. However, this is a special case and possible pertains to the node compartment (see below).

For C_1 to find its neighbors which can only be reached with the help of another compartment, it is necessary to inspect

- first, possible candidate compartments (C_2 in the previous definition) and
- second, possible entities E'_1 inside these candidate compartments

via which communication with C_1 neighbors (with respect to C_2) might be possible.

In principle, it would be possible to search for all possible neighbors by "broadcasting" to all entities in the same node compartment such a neighbor discovery request, which could then travel inside these candidate compartments C_2 to other entities of the originating compartment C_1, which could then decide whether to answer such a request or not. Depending on the number of adjacent compartments, on their size, and on the danger of cycles in such a discovery process, this might or might not be an efficient process.

To improve efficiency, a first measure is to impose a structural order on the set of all compartments intersecting with a node compartment and to restrict the propagation of such neighbor discoveries in the direction of this order. Simply put, we would enforce a layering structure! A more finer-grained control is possible, configuring inside a node compartment which compartments are allowed to search for neighbors over which other compartments; even control on entity level is easily conceivable.

Neighbor Discovery in a Node Compartment The previous paragraph discussed the explicit configuration of relationships of compartments inside a node compartment. This is often possible and desirable, but not yet the most general concept. In fact, we can take the similarities between "normal" compartments and node compartments one step further by also performing neighbor discovery inside the node compartment. In fact, a node compartment is special in the sense that it has an atomic notion of neighborhood; it is one of the few compartments which indeed have C neighbors without the help of other compartments.[4]

What constitutes a neighbor inside a node compartment? In principle, all entities inside a node compartment could be neighbors of each other. Since the cost of "broadcasting" inside a node compartment is negligible (compared to broadcasting in truly distributed compartments), one could simply decide to broadcast all kinds of such neighborhood searches inside the node compartment. In fact, this idea has, in the scope of the IP/TCP demultiplexing context, been proposed before by introducing a sort of "default" port between IP and TCP, with names for the TCP service included in such a search request for a TCP name [4].

In practice, only such entities that could provide a reasonable communication service need to be considered as neighbors. This observation gives rise to a *service graph* inside a node compartment. The service graph is a way of formalizing the relationship of the different data transport entities in that a link in this graph denotes the fact that a certain service (represented by a vertex) is required to build a higher service on top. As an example, a service described "ordered byte stream" as provided by TCP can build on top of a service "frame relay". The link in between the two would represent an entity that is *implementing* this service. Construction of such a graph requires (a) a certain language (an *ontology*) that describes transport services, (b) neighbor discovery on the set of entities, i.e., identifying which entities can render useful services to each other, and (c) a conventional link-state routing information exchange protocol that distributes the information about the existence of certain entities within the node compartment.

With such a semantically induced neighborhood relationship on a service graph, one entity can restrict the search for actual neighbors efficiently to those compartments that are able to render the required communication service in the first place. Note that in many cases the construction of this service graph will be a rather static

---

[4]The only other type of such compartments are those that directly represent physical communication, where neighborhood is realized by properties of physical signal propagation.

and manual procedure. However, it is described here in order to demonstrate that even the dynamic deployment of new entities (providing different services) can be handled without intervention of a central authority. All links departing from a certain node in the service graph represent entities that offer the required service and could be used to explore the neighborhood of an entity; this exploration is steered by so-called resource records (not discussed here further owing to lack of space).

5.2.4.6 Data Structures for Names and Addresses

To summarize the discussion about names and addresses, let us consider the necessary data structures. These are, so far, the *routing table* and the *resolution table*. Both tables exist, in principle, once per entity. As an optimization measure, a compartment can decide to implement these tables only once per compartment, per node instead of once per entity, and update these tables accordingly for all entities. Such a design decision trades off efficiency against more fine-grained behavior, e.g., per-entity source routing becomes rather hard to realize with shared routing tables.

An example of a routing table is shown in Table 5.1, reflecting the situation of Fig. 5.4. This table specifies the destination entity for which this entry is intended, the neighboring entity (in the *same* compartment) via which this routing takes place, and the cost of reaching the destination via this neighbor. It is a completely straightforward routing table and can be extended in the usual ways (e.g., storing routing paths instead of next hops). What is important to note is that it only stores identifiers of a *single* compartment. Note that neither the figure nor the table indicate which the assisting compartments (to realize neighborhood) nor the entities inside these assisting compartments are; this is immaterial for a routing process.

To relate this table to the reachability of a neighbor via different compartments, one might be tempted to add compartment names and names of entities in other compartment, but this would hamper information hiding as entity names should not be revealed outside of their compartment. To this end, it is useful to tie the routing table in with the resolution table, without explicitly mentioning any names of foreign compartments. This is the purpose of the fourth column of the routing table. Also note in the example that there are two *routing* options to get to entity F—via G and H—and that there are two neighboring relationships between E and G (with different costs) via the two references to the resolution table entries x and u. This gives entity E *three* different options to forward data to F.

The resolution table adds the missing information. It contains information via which assisting compartment a neighbor can be reached. Table 5.2 shows an example resolution table, some of whose entries correspond to the example shown in Fig. 5.5.

Note some of the structural properties of this resolution table: Each row contains information pertaining to two compartments—the one originating the need for a resolution of one of its names and another, assisting compartment providing the

Table 5.1 Routing table of an entity E in a compartment C_1

Destination entity	Via neighbor	Cost	Reference into resolution table
F	G	15	x
F	G	23	u
F	H	15	y
K	L	15	z

Fig. 5.4 Arrangement of entities inside a compartment for routing table example from the perspective of entity E. *Lines inside C_1* indicate neighborhood relation; *dashed lines* indicate indirect connectivity unknown to E. *Lines outside of C_1* indicate that neighborhood is realized by relying on some assisting compartment other than C_1

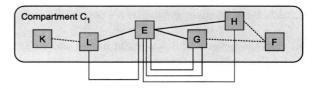

address bound to such a name. Moreover, there is a need to distinguish between the "local" entity and a remote entity in both compartments, with "local" being defined by the scope of the requesting entity's node compartment. The two entities in the originating compartment are obviously required as keys in this table (they define which local entity requested which name to resolve). The two entities in the assisting compartment are necessary to identify the actual remote address as the result of the resolution and the local entity in the assisting compartment via which this resolution had succeeded.

Let us look at this Table 5.2 a bit more closely. The first row, for example, tells how the name E_1 in compartment C_1 can be resolved into the address $N_{2.2}@C_2$. An alternative resolution for this neighbor N_1 of E_3 is given in the second row which uses another compartment C_4 (think, e.g., of the IP (C_1) address of a gateway, resolved via both an Ethernet compartment C_2 and a WLAN compartment C_4). The third row shows that the same originating entity can have resolutions for more than one peer entity; here, the resolution is provided via the same local entity in C_2. Finally, the fourth row shows the recursive structure of the resolution table as the assisting compartment C_2 and its entities/names of the first row assume the role of the originating in this row, needed resolution of its names itself via some other assisting compartment.

This resolution table is populated, as its name suggests, by the name resolution process. There is a large variety on possible options for name resolution, outlined already in Sect. 5.2.2. But the basic communality is always that a name valid in one compartment is resolved into an address from some other compartment. If such addresses can be found in some compartment, the name resolution has succeeded,

Table 5.2 Resolution table from the perspective of node compartment 2 of Fig. 5.5

Handle	Originating compartment	Local entity	Remote entity name	Assisting compartment	Local entity	Remote entity name	Cost
u	C_1	E	F	C_2	C	D	5
v	C_1	E	F	C_4	A	B	2
w	C_1	E	J	C_2	C	K	7
x	C_2	C	D	C_3	G	H	3

Fig. 5.5 Scenario for the resolution table example

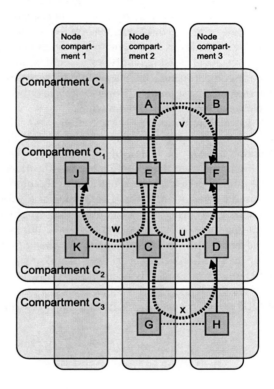

turning two entities of the originating compartment into neighbors. If name resolution fails, this does not mean that these entities cannot communicate—it only means that there is no single compartment available via which these two entities can be turned into neighbors. For them to communicate, they have to rely on routing and forwarding via entities in their own compartment, where each forwarding step might use different compartments along the way.[5]

---

[5]In such a case, there would be a least two different assisting compartments involved (else, they would be neighbors); it is conceivable, however, that the forwarding steps can all take place over the same assisting compartment and yet the two entities engaged in communication are not considered as neighbors because of access control limitations. This is, however, a rather specific case, as in

To emphasize this point again: Two entities E_1 and E_2 are considered neighbors only if there is an entry in the resolution table relating these two entities to each other. Else, E_1 and E_2 only can communicate with each other if they can do so via the compartment's routing procedure. This corresponds to standard practice: An IP entity knows, for a remote IP entity, only the remote IP name and the name of the next-hop IP entity (typically called a "gateway router"); it does not know and does not need to know a resolution of this name into a, say, Ethernet address.

There are two final observations to be made. First, the resolution table as presented here spans multiple compartments—not only a pair of compartments per row, but indeed possibly many compartment pairs. It is hence an implementation choice how to realize this table. One option would be to assign the respective rows to the local entities in the originating compartment and make them responsible for it; or to the local assisting entity; or it could be implemented as a data structure in the node compartment. The latter might be attractive as this table needs a carefully implemented access control. However, this is only an implementation choice, it has no relevance for our concept as a whole.

A second, more important observation is the relation to information hiding and the need to hide addresses from entities. This is indeed possible by assigning the resolution table to the node compartment and only putting pointers to rows of the resolution table into the routing table (as illustrated by Table 5.1). With a proper access control in place, it becomes effectively impossible for entities to find out addresses; the name of one entity is not related in any form to any kind of address it has in another entity. The only relationship that exists happens with row pointers from the routing table to the resolution table, and this is purely local information. As a consequence, some problems vanish (e.g., changing IP names when switching from WLAN to wired connection is not necessary) or become much simpler (e.g., switching IP names when mobility occurs). Much can be solved by a controlled change of these row pointers; in fact, this has been successfully implemented in some of the 4WARD prototypes and is described in the prototyping chapter of this book.

5.3 A Centerpiece: Name Resolution

5.3.1 Name Resolution Is Neighbor Discovery

From the discussion above, name resolution is the process of learning about the mapping of a name N_1 of an entity E_1 in a compartment C_1 to a name N_2 of another entity E_2 in another compartment C_2, where E_1 and E_2 share the same node compartment. In addition, it is also clear that such a mapping is only of relevance to an entity E that also belongs to compartment C_1; other entities do not even understand the semantics of this namespace (or cannot communicate with E_1).

---

fact this would make the membership of the two communicating entities to the same originating compartment rather questionable.

Hence, there are two design options for a name resolution system: One is to make such a mapping known inside compartment C_1, even to entities E in C_1 that have no means of accessing C_2. But for such an entity, the address $N_2@C_2$ is meaningless and useless. The alternative is to provide such a name resolution result only to such E in C_1 that can indeed access C_2. If that is the case, then E and E_1 are neighbors from the perspective of C_1, by virtue of being able to access C_2.

Hence, a successful name resolution always means that a neighbor has been discovered. Put briefly:

Name resolution is always neighbor discovery.

This perspective might seem to be at odds with current practice. Name resolution systems do resolve names into addresses even if the originator of the name resolution request and the target do not appear to be neighbors—for example, resolving a DNS name into an IP address. This is, however, simply a misconception: From the perspective of the entity issuing the name resolution request, itself and the so-named entity indeed are neighbors, facilitated by the IP layer. The IP layer (in its simplest form, ignoring issues like firewalls and middleboxes for now) creates the abstraction of a fully connected graph, turning all entities using it into neighbors.[6] Whether or not such an abstraction is practical highly depends on the semantics of the compartments sitting on top of it, as well as on technical and technological circumstances (e.g., size of the assisting compartment).

5.3.2 Discovering All Neighbors as a Special Case

So far, we regarded name resolution as a special case of neighbor discovery—finding out whether a particular name is owned by a neighboring entity. The vice versa view is just as valid: neighbor discovery is a special case of name resolution. We simply need to identify, for a given compartment, a "wildcard" name, to which all entities of this compartment answer when asked whether they carry this name (and include their actual name into the answer). Details about how such a wildcard should look like, whether it has a restricted semantics of some sort, whether indeed all nodes would answer, etc. are all design decisions taken by a specific compartment. Nonetheless, a wide range of choices fit into the conceptual framework presented so far.

---

[6]This does of course not mean that all IP entities themselves are neighbors. Clearly, the IP entities themselves are not all neighbors; rather, any entities E_1 and E_2 belonging to some compartment C and bot connected to IP are considered C neighbors with respect to IP, according to the definition of Sect. 5.2.4.4.

5.3.3 Name Resolution vs. Routing

Based on neighbor discovery by means of name resolution, an entity finds out its
neighborhood and knows how, via which compartment, to reach its neighbors. Based
on neighborhood information, routing tables and, thence, forwarding tables can be
constructed. As such, the interaction of name resolution and routing is straightforward.

What does this entail, however, for entities that do not wish to engage explicitly
in a routing protocol (say, the IP engine of an end device)? There are two conceptual
approaches to solve this:

- Mandate that, indeed, each entity must participate in the routing protocol of the
  compartment. Neighboring nodes that do act as routers would then offer routes
  to the actual destination. The routing protocol implementation can be limited or
  simplified, depending on the roles an entity assumes inside a compartment. Any
  notion of cost is naturally expressed by the routing protocol.
- Extend the semantics of neighbor discovery/name resolution. Put simply, have
  an entity answer not only to name resolution requests for which it does own the
  name, but also to requests for names for which it does know a route. Such an
  entity would pretend to own a name, when in fact it would only forward data
  toward the destination; the requesting entity is under the misconception of being
  a neighbor of the intended entity. To properly reflect costs of forwarding, the costs
  of communicating with the (pretending) entity via the assisting compartment must
  be properly modified.

The first approach is the architecturally much cleaner solution. The second one
can be acceptable in some circumstances, and in fact is frequently applied (e.g.,
HTTP proxies over IP).

5.3.4 Configuring Name Resolution

The discussion so far has implicitly assumed fairly simply name resolution scenarios: an entity of one compartment seeks to resolve a name (possibly, a wildcard)
over another, assisting compartment. How such a request is distributed in the assisting compartment has not been considered. We need to close this gap now.

5.3.4.1 Two-Compartment Situations

Broadcast-Based Resolution The basic case consists of just two compartments:
a compartment C_1 where the name resolution request originates and one assisting
compartment C_2, from which the address of the entity should be provided.

In the simplest case, such a resolution request is simply broadcast in the assisting compartment. An entity E in C_2 receiving such a request can detect for which

originating compartment the request is intended. This entity has, however, no means of deciding which, if any, of those entities in the same node compartment as E *and* in C_1 carries the desired name—recall that names have no meaning for entities not belonging to the proper namespace. Hence, the only option for E is to distribute the request to all local (= in the same node compartment) entities in C_1. This is a broadcast within the scope of a node compartment and hence quite limited; the first broadcast inside C_2 is the potentially costly one. Note that in this scheme, there is no need for any entity to register a name with anybody (whether or not an admission control and name allocation protocol is executed by the compartment beforehand is an orthogonal question). This setup immediately reflects typical solutions like ARP and IP.

Obviously, an entity in C_1 can seek name resolution over multiple compartments $C_2, C_3, \ldots$ in parallel or sequentially; it is not limited to any single one compartment.

Lookup-Based Resolution   If the cost of broadcasting in assisting compartment(s) is deemed prohibitive, the compartment C_1 originating the names has to provide some additional structure (no assisting compartment can help for lack of semantic understanding of the names of C_1).

The simplest structure is to designate an entity E_R inside C_1 as a repository for name/address mappings. Any entity inside C_1, upon binding a name to an address inside some other compartment C_2, would inform the repository of such a binding. When seeking a resolution, an entity contacts the repository and asks for the corresponding address. This is simple to do for the repository as it understands the semantics of names from C_1 (being a member) and can hence detect the correct bindings. With obvious techniques, such a repository can be replicated, distributed, or turned into a hierarchical structure.

The possible pitfall lies, however, in "contact the repository". For this to work, the requesting entity has to know (a) the repository's name inside C_1 (which is simple) and (b) an address of the repository—else, how to tell an assisting compartment where to transport the request—or a neighbor that can forward to the repository. In many cases, such an address inside a widely available assisting compartment can be considered to be well known, can be pre-distributed during deployment, or can be configured during the admission control procedure when joining a compartment. A typical example for the later is DNS as a compartment knowing about the namespace of fully qualified domain names and the IP address of a DNS name server is distributed as part of a general network configuration process like DHCP (mixing up the joining of different compartments in an architecturally unclean manner).

More complex lookup structures are conceivable as well. Instead of assuming that an address or a route toward such a repository is known, a request could be sent on a random walk through a compartment, or it could be broadcast inside the requesting compartment (as opposed to broadcasting inside the assisting compartment, as discussed above), or gradient techniques (e.g., for location-based networking) could be used. Many options exist here, and they have been mostly investigated in the context of mobile ad hoc networks or wireless sensor networks; in

Table 5.3 Resolution configuration table

Originator compartment	Resolver name	Assisting compartment	Resolver name	Helper compartment	Helper name
IP	*	Ethernet	bcast	–	–
DNS	DNSResolver	IP	1.2.3.4	–	–
Music	–	IP	–	P2P	P2PEntryHost
P2P	P2PEntryHost	IP	5.6.7.8	–	

standard tethered or cellular networks, mostly conventional solutions are deployed today.

5.3.4.2 Name Resolution with a Helper Compartment

If the procedure to find a name/address mapping turns out to be rather complex, it becomes attractive to "outsource" it into another functional subsystem. For example, consider a peer-to-peer storage based scheme to hold these mappings. On the one hand, it is conceivable and semantically advantageous to integrate such a peer-to-peer system into a given compartment. On the other hand, a generally available peer-to-peer system to store such bindings might be useful for many other compartments. The disadvantage of such a third, "helper" compartment (as opposed to the compartment originating the name and the assisting compartment providing the address) is that the semantic understanding of names is lost and that only simplistic tests for equality of names can be applied. Whether this is desirable and sufficient highly depends on the concrete needs of the originating compartments and the semantic complexity of its namespace.

5.3.4.3 Name Resolution Configuration Table

These options have to be made known to an entity before it can initiate a name lookup in the first place; similarly, an entity has to know if and where to register its name/address bindings. Many operating systems have a notion of a "name resolution configuration"—we generalize this concept here and introduce the *resolution configuration table* as shown by example in Table 5.3.

The first row of this table shows a typical IP and ARP setup: When resolving an IP name, send a request to all IP entities reachable via the Ethernet compartment, using the Ethernet address "bcast" in this request.

The second row is the also typical example of resolving fully qualified domain names (designated as "DNS" namespace) into IP addresses. This is more complex than ARP as there is a specific entity in the DNS compartment, here called DNSResolver and corresponding to a DNS name server, for which the address in the assisting compartment is already supplied in this resolution table (doing a broadcast search in the IP comportment is likely not efficient).

The third row is the most complex one. It assumes a "Music" compartment whose names shall be resolved into IP addresses. But the Music compartment as such does not provide such resolution functions but rather relies on the "P2P" compartment for doing so; in this compartment, the entity named "P2PEntryHost" shall be contacted for Music resolution requests. How to find this name is explained in the fourth row. This fourth entry is analogous to the second entry.

The entries of this resolution configuration table are relevant to the originating compartment and can, hence, be considered to belong, per row, to them. For convenience, we have only shown a single such table here.

With this table, our list of tables is almost complement. We have the (a) routing table, per entity or per compartment, (b) the resolution table, per entity or per node compartment, and (c) the resolution configuration table, per node compartment.

5.3.4.4 Bootstrapping the Name Resolution Configuration Table

A final remark is in order about the bootstrapping of this configuration table. Two columns in this table are indispensable: the originating compartment for any resolution attempt, as well as the name under which the resolver is reachable. All the remaining columns, however, can be regarded as hints and do not have to provided up front to a node (in practice, they typically would be provided for efficiency). For example, in absence of any information how to reach "DNSResolver", all that is to be done is to solve a routing problem in the DNS compartment. This means, it is necessary to run neighbor discovery in the DNS compartment, finding possible neighbors via, e.g., IP, and then run a routing protocol to figure out how to reach DNSResolver. In this sense, there is no magic information involved in this configuration table; if all else fails, even the content of the resolution configuration table can be automatically generated by falling back to primitive search methods.

5.3.5 Late Resolution as a Special Case

The usual notion of name resolution is that of obtaining the mapping of a name onto an address *before* the actual communication is initiated, before the assisting compartment is asked to forward the first packet toward the destination address. While the previous discussion has followed this line of thought, there is nothing inherent in our framework that mandates such *early resolution*.

Rather, it is just as well possible to realize *late resolution*. Instead of completing the resolution process before the first packet is forwarded, resolution and forwarding can be linked with each other. A first resolution step might not provide the final destination address, but only the address of an entity (inside the originating compartment) that can provide more information about the destination address. This process iteratively refines the resolution accuracy, until eventually the actual destination address is provided.

In a sense, a P2P resolution process follows this scheme anyways. What is more interesting is to organize the resolution process such that the forwarding from one resolving entity to the next is guaranteed to reduce the distance (measured in the assisting compartment, not in the originating one) to the actual destination address as well, despite the lack of precise knowledge of the destination address. How to solve this challenge in detail is, as of today, a matter of active research.

To outline some of the current ideas on this topic: Suppose a compartment C_A uses a hierarchical namespace, where the hierarchical namespace maps to a hierarchical routing structure. Suppose we have a further compartment C_R with a flat namespace, which intends to use names of C_A as addresses for its own entities (A for "assisting", R for "requesting").

In an early resolution scheme, a name in C_R would simply be resolved into the full address out of C_A; the name resolution system of C_R would simply store the full binding. After this full address has been looked up, C_A is free to act on this address as it chooses, e.g., by doing a hierarchical routing/forwarding on it. This is common practice.

In a late resolution scheme, names of C_R would not be resolved into a complete name in C_A directly; a name would not have a complete address. The interesting design challenge here is to ensure that the two namespaces stay semantically separate, that there is no need for one of the compartments to understand the other's name structure—else, introducing new namespaces quickly turns into a nightmare. One option to realize this requirement is given by the name registration: When an entity in C_R seeks an address from C_A for one of its entities, it is given a bit string, to be stored in the name resolution system of C_R. While this bit string is opaque from the perspective of C_R, it carries meaning inside C_A (these are names!) and the meaning could be a *list* of prefixes of the actual address, from shortest prefix to full address (we assumed a hierarchical namespace in C_A). When an entity in C_R resolves the desired name, it obtains the opaque bit string, passes this to an entity in C_A, which realizes that it has been given a hierarchical list of address prefixes. Depending on its own position in the topology of C_A with respect to this list, it can choose to forward to the destination directly or to consult a proper *name resolver of C_R* again, at a place closer to the actual destination. This would give considerable freedom in the interaction of name resolution and routing, for the price of more complex address lists to be sent around. In many cases, though, a simple hierarchical routing scheme would be sufficient.

5.4 Conclusions

This section has discussed some basic ideas about names and addresses, their relationship to each other, their relationship to structural properties of a communication system, and the basic data structures needed to tie them together. We hope that we have contributed to the clarification of terminology and understanding of these fundamental notions in communication systems.

In particular, we have identified five crucial data structures: (a) a binding table describing which entities have bound themselves to names of other entities (typically in other compartments), (b) a routing & forwarding table that exists per entity or, as a simplification, per compartment and per node compartment, (c) the name resolution table, which describes how two neighboring entities of one compartment *actually* communicate with each other using which entities in which other compartments, (d) the idea of a service graph, which contains all *possible* usage relationships of entities within a node compartment, and (e) a configuration table for the name resolution process, describing which compartments can attempt name resolution via which other compartment and what parameters are required to do so (with service graph and name resolution configuration closely tied in with each other).

Based on this abstract treatment, we have discussed the process of name resolution in a general fashion. Many communication primitives can be cast into this light, for example, we came to realize that neighbor discovery and name resolution are, at the core, the same thing. We believe that a rigorous treatment of communication design that gives name resolution its proper place across the protocol stacks will lead to a more general, more extensible, and more flexible communication system than what we are currently faced with. It can incorporate automatic discovery of communication opportunities and can solve problems like session mobility quite easily. From a practical perspective, it also can handle a multitude of namespaces and protocol families, without having to standardize on any single namespace across widely differing communication needs.

References

1. G. Coulouris, J. Dollimore, T. Kindberg, *Distributed Systems Concepts and Design*, 4th edn. (Addison Wesley, Reading, 2005)
2. J. Day, *Patterns in Network Architecture—A Return to Fundamentals*, 1st edn. (Prentice Hall, New York, 2007)
3. C. Jelger, C. Tschudin, S. Schmid, G. Leduc, Basic abstractions for an autonomic network architecture, in *World of Wireless, Mobile and Multimedia Networks, WoWMoM 2007, IEEE Intl. Symposium* (2007), p. 16, http://scholar.google.com/scholar?hl=en&btnG=Search&q=intitle:Basic+Abstractions+for+an+Autonomic+Network+Architecture#0
4. M. Lottor, RFC 1078: TCP Port Service Multiplexer (TCPMUX) (1988), http://www.ietf.org/rfc/rfc1078.txt
5. S.e. Randriamasy, Mechanisms for Generic Paths (2009)
6. J. Saltzer, On the Naming and Binding of Network Destinations (1993)
7. J.F. Shoch, A note on inter-network naming, addressing, and routing (1978), http://ana-3.lcs.mit.edu/~jnc/tech/ien/ien19.txt
8. N. Shroff, R. Srikant, A tutorial on cross-layer optimization in wireless networks, IEEE J. Sel. Areas Commun. **24**(8), 1452–1463 (2006), doi:10.1109/JSAC.2006.879351

Chapter 6
Security Aspects and Principles

Göran Schultz

Abstract Rethinking the fundamental network architecture seems to be able to solve some known architectural security problems of the existing internet, but proposals are also investigated more thoroughly from the security angle overall. The information-centric approach of 4WARD is built on the concept of securing information rather than locations and paths used for information transit. Doing so, the security principles based on ownership and controlling access at the originating source become challenged. At the same time, moving intelligence into the network itself challenges the underlying assumption of having an Internet consisting of neutral, dumb, and fundamentally cooperating and trusting autonomous domains. 4WARD states the security principles necessary for dynamical management of virtualized, largely self-configuring entities having specific properties. The specific security implementation choices necessary for network design, transport, routing, lookup, privacy, accountability, caching and monitoring are part of the design process, for which 4WARD contributes functional descriptions and the concept of a design repository. 4WARD acknowledges and considers the business and governmental control interests that will heavily influence the security direction into which the future network evolves.

6.1 Introduction

Listing security principles ranging from access and availability to privacy and non-repudiation for the Future Internet does not by itself give meaningful insight into problems faced by an ever-changing information society. The 4WARD architecture framework and its networking propositions deal both with an abstract concept of

G. Schultz (✉)
Ericsson Research, Jorvas, Finland
e-mail: goran.schultz@ericsson.com

L.M. Correia et al. (eds.), *Architecture and Design for the Future Internet*,
Signals and Communication Technology,
DOI 10.1007/978-90-481-9346-2_6, © Springer Science+Business Media B.V. 2011

handling information for the benefit of end users, as well as with the conflicting interests arising from a multitude of players with vested interests—governments wanting control over information flow, operators and content providers positioning themselves as something else than bit movers, and non-commercial demands for privacy and accountability, spiced with technical, operational and legal limitations on what can be achieved.

There is a realization that the existing Internet took the world by surprise, growing without constraints and nurtured by the best of skills and intentions. As the business world and general public started to dominate usage, misuse and ownership issues started dominating the headlines. A younger generation has grown up with concepts of freedom (as in free beer and free speech) and a feeling that only a multitude of sources of information can replace the fictitious correctness and reliability of a single trusted source, with the drawback of no filtering that comes with traditional publishing. The possibility that the underlying network itself cannot be trusted is sinking in slowly if at all. The learning curve regarding privacy lost due to social networks seems to work much faster, but the trail of crumbs left behind on the Internet has many security aspects due to the ease by which terabytes of stored data can be combined. The private security-industrial complex that arose after the events of September 11, 2001 is in the USA able to bypass the constraints on government search [30], and cloud computing will make this even worse. In several countries, a new trend is to force internet users and content publishers to register with their real identity.

In the 4WARD discussions leading to the architectural framework, there has been two security tracks. For information centric networks, the abstract concept of information itself carrying the necessary pieces to ensure integrity is combined with the ideas of publish/subscribe, i.e. giving the receiving party control over what comes down his line. The other track deals with the burden of dynamically managing the network, in particular how to provide particular characteristics, e.g. QoS, to a secure and scalable user-invisible infrastructure, where self-management and self-configuration places difficult security requirements on parties that traditionally have been reluctant to share business-related resource information.

4WARD work is properly set in relation to other similar ongoing future network efforts. The GENI [10] effort in the US explores software-defined networking allowing operator-users the ability to deeply program the network devices they utilize. The global routing problem is addressed within a related Floating Cloud architecture of the overlay type, with a testbed utilizing MPLS to bypass the current routing protocols, allowing the tradeoff between economy of bunching versus granularity for control and management for security reasons. The wireless networks in use are characterized by intermittent disconnections forcing studies of delay-tolerant networking, bringing into reconsideration the security associations used—the trust relations used for social relations might be an alternative to a fixed DNS for name resolution. At PARC [25], the content-centric aspect for delivery over whatever network happens to be available is one possibility for an information-centric future network, with security traced back to sources, but content having multiple locations.

As a contrast to the internet way of dealing with security problems, there is the rather successful cellular phone system approach of having a tightly operator-controlled authentication system combined with a clear separation between network signalling and user data traffic. Approaches to use this authentication, e.g. using OpenID, are discussed later in this chapter.

- Integrity: nobody has tampered with content
- Confidentiality: only those who should see the content can do so
- Authenticity: the content is what you think it is (note authentication as a process, for information and users)
- Availability: access to content or networks is not blocked for auxiliary reasons
- Authorization: access to content based on verified identity in a broad sense
- Non-repudiation: you can't deny that it came from you
- Trust: understanding the fine granularity of access and transactions, containing elements of time, location, accumulated history, economics and conflicting interest
- Accountability: post mortem possibilities of analyzing and taking legal and technical action
- Location: geographical information about users and information caches and handlers in the logical topology

In conclusion, future networks will have challenges both when it comes to locating information and to getting hold of it. The signalling needed to locate information has similarity to and security problems identifiable already in the existing Internet. The actual delivery of bulk content will have multiple solutions many of which are strictly a result of business considerations.

6.2 Business Models and Security Implications

6.2.1 Owning as a Concept in the Digital World

Security means very different things to different actors, and the paradigm changes introduced by future internet architectures affect the basic building assumptions in several ways. In one dimension, security still remains a chain with failure at the weakest link destroying the whole structure. In another dimension, understanding the underlying technological vulnerabilities allows us to focus on aspects that matter in the end—the usability of technical solutions.

- Is there a usability mismatch between users, technology and content value?

A vexing problem is the reluctance to demand and pay for security, which does not become easier when more niche players than before are needed for a working totality. The simplicity of usage provided by single sign on has had a dark side of scramble for control over user identity, and the reasonable desire to provide differentiated services has had an ugly aspect of deep packet inspection, everything happening outside any general legal framework or border-crossing operational understanding of what is reasonable and fair.

- Who pays for security?
- Where does an entity get an identity; who "owns" the user?

The existing Internet has an "intelligence at the edge" structure, which allows end to end security for packets, but trusts the DNS and transport infrastructure to work correctly or not at all. Adding mobility forced rethinking of the dynamic security, in particular, the return routability solutions of IPv6 highlighted a larger problem, i.e. the reliability of the bindings in the infrastructure. Spoofed MAC addresses on the link layer and incorrect DNS or routing table entries require analysis of availability, an often forgotten aspect of security.

On the lowest level, security needs to be addressed on verification of the hardware and software of the nodes making up the system.

On the following level, security becomes a management problem when dynamically building a network from the functional blocks provided by the 4WARD design repository. The security anchors optionally required by management interfaces (SSPs) and data interfaces (SGPs) are use case specific, and will be partly provided when downloading the service layer specifications in the self-configuration scenario envisioned, partly provided as certificates included in the handling of information objects.

Monitoring adherence to a specific service level agreement is loaded with business and security considerations. The business interest of data mining of traffic is in conflict with user privacy, and the view of traffic patterns gives network competitors insight into the resources and capabilities that are commonly withheld for business reasons. There is reason to believe that the future internet needs extensive dynamic information sharing between autonomous parties on a scale unheard of in the existing network, the alternative being stupid bits pipes with prescribed capabilities, which scenario the operators also wish to avoid. When sharing information, a method for sharing revenue becomes necessary, and a general framework for fine-grained trust among cooperating networks becomes necessary—the bilateral model for service exchange is outdated, and new methods for whitelisting and blacklisting autonomous systems are around the corner.

- Service level agreements will contain policies, monitoring, logging and trust on some level of granularity

At the highest level, information has been detached from a particular node, and only the distinction between "owned" information or "self-certified" information becomes relevant. The privacy requirements demand that ownership can be expressed as a pseudoidentifier, which then needs to be anchored at a trusted third party to the actual user identity. The 4WARD approach to naming allows self-certified ownership to be attached to anybody who possesses the corresponding secret key of an information object.

- Ownership is defined in terms of possessing a private key
- Pseudoidentifiers give privacy with accountability

From the neighboring autonomous system's point of view, caching and storing unknown information has to have some economic incentive, since useless retransmission might generate revenue, while caching only allows bandwidth saving. Accountability moves from the current hosts to the identifiers and networks that verify the identifiers, from the point of view of including an entry in a name resolution service. The current internet WHOIS [36] identification of domains must take on a new role in future networks—right now, a public registration of a domain (which helps targeted attacks) might be falsely identified, while a private registration shifts the legal responsibility onto the domain name registrar.

The price of caching is coming down much faster than the price of transmission, so overall the picture favors the 4WARD information-centric model. The legal responsibility for storing "child pornography" as cached, possibly encrypted bits, is uncharted territory. The focus instead shifts toward the name resolution system—if you cannot access information, it is meaningless even if it exists somewhere. Operators are likely to fight for their own slice of the future internet name space—in this case, the "human understandable" way of addressing information that perhaps through a dictionary search, using some kind of ontology language, e.g. OWL [35], maps to 4WARD NetInf information objects. While a simple flat name space will not be global for search delay reasons, having an architecturally mandated hierarchy spoils the purpose of introducing the new name space. The need to be able to assign names that are not globally unique must be balanced against the possibility of "spoofed" sabotage of existing named concepts.

The availability implications of moving information away from the owner's host as happens in cloud computing come into full view when law enforcement impounds a server hosting a large number of businesses—and causing a full stop for all, due to one company being investigated.

The legal framework for a society depending on shared resources is wanting. The privacy aspects of data stored in a shared facility where somebody else than the data owner handles the security keys raises fundamental questions of trust and responsibility, especially if the physical and judicial location of data and data management is different.

6.2.2 Life in a Goldfish Bowl

Data-mining, however useful for googling information, is rapidly becoming the plague of internet users. Lip service to privacy on governmental level [34] is not worth much without competent enforcement. The SafeHarbor agreement between the US and the EU has a ten year track record, and a detailed review [33] shows the glaring conflict between having a logo on a home page, and actually adhering to the text it represents.

The privacy problem regarding usage in digital technologies has three nontechnical aspects. While a piece of information might be public, the new aspect is the ease (low transaction cost) by which it can be accessed in volume and combined with information from other sources. The second aspect concerns accountability—a physical library might require identification from users that enables tracing of use afterwards. Only with respect to healthcare records have such logging requirements gained acceptance on the digital side. The third aspect is a fundamental disagreement on ownership of personal data residing in a repository, with the EU giving weight to the individual concerned, at least in theory.

• Worst case scenario is the present case of lack of understanding of privacy failure

The privacy problems for future networks are linked to accountability. Legal interception and logging of internet access is unavoidable, but the technical ease of doing so should be controllable. IETF has a documented view on wiretapping in general [12].

One possible step toward privacy was the 2007 W3C machine readable protocol P3P [26] that in .xml defines what the requesting party expects from the privacy point of view. The obvious problem is enforcement, and here new efforts are under way [9]. OAuth [22] is a somewhat debated approach of authorizing without giving up privacy, development effort that is now moving from OASIS toward IETF. OAuth should be seen as a counterpart to OpenID [24], which is an attempt to define identity without the lock-in of managed name spaces.

The 4WARD architecture strives to avoid centralized structures. For anchoring security, the options are limited but the technical implementation has a variety of possibilities. For users, the ultimate anchoring of identity could take place at a 3G operator as happens today with SIM/USIM cards, leaving open the hairy question of how well a person has been identified when a card is issued. Attempts to have ultimate anchoring done by the government ("police") have not met with great practical success, e.g. the Finnish identity card. A promising development is the usage of ephemeral identifiers tied by cryptographic certificates to the 3G HSS system. These can be linked to identities provided by nonprofit identification organizations such as OpenID.

For information detached from a host, some relation to the originator must be maintained in the metadata by which the information itself is located. For privacy reasons, users should not be identified directly, instead (rapidly) changing pseudo-identifiers meet both privacy and accountability requirements. Such methods are already in use in 3G USIM systems for security purposes, although not for privacy reasons.

The user identification enters the information centric network through ownership of secret keys corresponding to the public keys needed for verifying authenticity. While the public key has to vary for privacy reasons, the proven ownership of a public key amounts to an identity that can be held accountable when technically or legally needed. In practice, there is no need to check certificates back to the issuers except when issuing new certificates in the certificate chain. An act of publishing information corresponds to such an event; for subscribing/consuming information, checking the last certificate issued by the local delivering dictionary system and the certificate presented by the subscriber might suffice. The security controller needs to be part of the storage (caching) system—information exists if and only if you have access to it.

On the technical level, privacy requires additional consideration. Changing the pseudo-identifier is worthless if MAC numbers used in transit or public keys used for cryptographic verification gives away the identity to a listener. The actual list of considerations is longer. Network design and operation should avoid transport solutions that enable data-mining beyond legal requirements.

- Adding accountability solutions without addressing the privacy dimension will exacerbate the current problem
- Usability is the stumbling block for security
- Limiting damage and managing faults and changes is a pragmatic replacement for perfect solutions

The social networking tool Facebook has faced a privacy challenge by Canadian authorities on multiple counts. The globally interesting aspect is that Facebook apparently shares all personal information of its users with a million companies that operate third-party applications in 180 countries. The eBay Inc. corporate family has its servers in the US, which means that despite visible local presence in Europe, ebay.xx, the privacy laws applicable will be the ones valid in the US. For privacy issues eBay refers the users to TRUSTe, which not surprisingly makes a privacy reference to Safeharbor [34] mentioned earlier. Users desiring privacy in addition to security face the problem of not sticking out—possibly the real reason for security agencies making a tool such as the onion router [23] generally available. For true privacy to avoid timing correlations, the link up to the first onion server should always be fully utilized (by sending junk traffic), but in practice this part of the arrangement might not be practical. Some new technologies are providing true security but are possibly stumbling on the unavoidable requirement of accountability. Phil Zimmermann's Zfone [37] handles the difficult key management issue for encryption of VoIP traffic, but the proposed internet protocol ZRTP also effectively prevents legal interception. Keybased encryption allowing keys to genuinely self-destruct over time in P2P file systems has been developed [18]—while the encrypted data once distributed onto the network cannot be recalled, it is rendered useless since nobody can recreate the necessary key that was scattered across the file sharing system. Finding a balance between technological solutions and usable security for future networks depends on the competence level of the actors involved. For building the core networks, only the highest levels of technical security can be acceptable. There is no technical solution to network owners deliberately cheating, so the economic incentives to play fair have to be in place, supported by border-crossing legal support. For users both consuming and providing content, the security aspect becomes more complicated. There is no reason to expect an honest user not working in information technology to understand the details asked in pop-ups commonly used today. Instead, we need to build in tracing and verification into both software and hardware, with the aim of punishing deliberate abuse instead of focusing on the technology used to commit illegal or unethical acts.

6.2.3 Managing Security and Secure Management

The internet consists of loosely connected networks, each of which is running its own rules, to a certain extent. The common aspect joining the networks has been the IP stack, defining a name space with common rules. On top of this IP name space lives a human-understandable URL name space also controlled by ICANN [11], and a translation between the two done by the DNS domain name system (e.g. BIND). When looking at the network as a set of horizontal layers, a name on one layer gets mapped onto an address on the next, in a recursive fashion—address becomes a name for the next layer [5], as is discussed in more detail in the preceding chapter.

Future networks are possibly abandoning IP as the common waist of the communication stack. The local "domain" under single ownership, be it virtual or physical, can be secured by anchoring the components by local management. Conceptually, slices of resources can be allocated to users allowing building of new types of networks, with user-controlled routing and transport elements. The Stanford OpenFlow protocol [32] is an example of how this can be implemented. Here, an important aspect is to place security into an external controller, making high-level decisions about access control.

Abstracting beyond each particular implementation, security for network design has to have anchor points. For scalability reasons, trust based on pairwise-established relations between networking parties is problematic. 4WARD principles for trust handling can be implemented with trust chains, or more specifically, using certificates that allow fine-graining of trust, e.g. SPKI [31] and KeyNote [13]. A certificate contains a hierarchy of certificates and rules for applicability, as well as delegation rights and time to live. Such ideas are currently developed within the EU PSIRP [27] and RIT Floating Cloud [28] projects, allowing flexible tradeoff between efficiency and granular control.

In conclusion, the necessary security anchor points will take several forms. On the device level, a specific physical component is used as a trust anchor. For single ownership structures, the rules applicable are downloaded to nodes at boot time, according to 4WARD In-Network Management principles. For exchanges between networks, the basis is trust hierarchies allowing flexible and verifiable dynamic usage of resources.

6.3 Security Aspects Pertaining to the 4WARD Architecture Pillars

6.3.1 Virtualization of the Physical Substrate

Network Virtualization in 4WARD aims at sharing a common physical architecture among several business players. The **Infrastructure Provider** creates virtual resources over the physical infrastructure he owns and leases them to **Virtual Network Providers** (VNet Providers). The VNet Providers use these resources to build virtual networks (VNets) which are leased to **Virtual Network Operators**. The VNet Operator runs a VNet and provides traditional and new networking services to the end users.

The physical resources may then be shared by business competitors. Robust isolation between the resources they use is fundamental. The main security requirements to meet in order to guaranty isolation are:

a. Infrastructure Provider needs to trust all OSs and applications running on the physical node including VNet Provider applications.
b. VNet Provider needs to trust the physical node which may be mono or multi VNet.
c. The physical node must provide trust isolation capabilities between VNet Providers.
d. VNet Provider should be able to store and manage his own cryptographic keys in such a way that they are isolated from the Infrastructure Provider and from other VNet Providers.
e. VNet Provider should be able to update his OS and applications securely while running. Neither the Infrastructure Provider nor other VNet Providers should have this capability regarding the software they do not own.

The approach proposed in 4WARD to meet these requirements is derived from TCG specifications. TCG specifications aim essentially at providing means to secure Digital Right Management applications over personal computers.

According to the current knowledge, it's likely impossible to prevent a malicious code in a system without any hardware assistance [29]. This is why a specific physical component is used as a trust anchor. It contains an immutable code that is executed when the physical node is powered on. It also stores sensitive data such as cryptographic keys, hashes, etc.

- Booting up hardware and software has to have a trust seed as anchor point

The modern computer is composed by a complex set of hardware and software components. If one of them is corrupted, it may compromise the security of the whole system. In an environment where the trust is essential such as the network nodes described in network virtualization, it is necessary to verify the integrity of each component before activating it. When the system is powered up the first component is called "Boot Process". This component needs to be trusted. It is the trust seed, and it constitutes the anchor point for the trust of the entire system. The Boot Process has a task of verifying the integrity of the hardware and the next software to be activated. A chain based on this principle is built.

6.3.2 Building Paths—The 4WARD Generic Path

The 4WARD Generic Path highlights an addressing problem already present in the existing internet. The nodes used as part of the transport infrastructure have to be reliably identifiable. While the management of the different GPs is up to the GP designer, the actual management logic comes from the 4WARD INM, deciding what aspects are exposed.

An Entity has a name in a compartment but it has to be registered by a trusted authority. In P2P, a index server or a DHT is used, but this is not trustable and open to many security issues. For lower layers, like Ethernet, the name of the entity (the MAC address) is registered in a data base maintained by the operator.

Entities, which have names, are bound to other entities with names, which serve as addresses for the former. These names/addresses have in some circumstances to be routable in a routing plane. The name/address resolution is currently done via DNS in the Internet, but DNS servers are very weak points in the architecture.

When an entity wants to establish a Generic Path (GP) with a peer entity, one possibility is to introduce a negotiation phase through a server performing some mediation between entities. This server could be a rendez-vous point as in DONA. It has been envisaged to be able to join and leave a GP. One needs to specify a protocol for joining a GP in a trusted way. The GP should thus have some embedded security functions to control its endpoints. Such a GP could be a swarm or a multicast GP.

4WARD considers the use of disparate capability connections directly as parts of the Generic Path toolkit. WLAN, RFID, Bluetooth and Zigbee devices might conceptually form part of the future networks. Mobile telephones of the GSM and 3G variety are currently pasted onto the IP name structure with a one-to-one DNS translation. The problematic growth of routing tables originates from multihoming, where customers and ISPs like to be independent of address prefixes from providers in inner tiers. The new competing internet telephone technologies such as Skype utilize only the IP transport structure, having their own proprietary internal name spaces and translation tables, e.g. Skype-In and Skype-Out. The hierarchical structure of IP avoids loops but requires a central naming authority ICANN. A specific security problem arises today with the DNS servers that have come under attack [1]. A global effort is under way to convert all DNS servers to use proper authentication [7, 15–17] when accepting changes.

6.3.3 Network of Information

The Distributed Hash Tables (DHT) are suitable for flat name spaces such as cryptographic hashes, after having resolved the human-understandable semantics. However, a global name space built on DHT is not realistic due to search delays. Using hierarchical DHT methods such as Chord [2] allows building scope into the naming system—an organization might want absolute guarantees of information not leaking out beyond its own network.

The 4WARD naming scheme contains a tag which identifies how the full information object name should be interpreted. The tag allows changing the encryption algorithms and the introduction of secondary names for objects in the metadata, among other things.

For finding information that is floating around globally, a naming scheme containing the originating network is suitable. The Data Oriented Network Architecture (DONA) [3] naming approach consisting of two parts, the network part P and the locally meaningful label L, solves this. In the DONA approach, the P part is cryptographically related to a public key identifying the network, and settles part of the security concerns in this respect.

To find information in general, the transition from human-understandable to scalable machine-usable naming contains security pitfalls. In the existing internet, abbreviated URLs have opened new opportunities for malicious redirection. The Content Centric Networking (CCN) [25] promoted by Van Jacobson assumes the existence of an IP-like hierarchical structure for information objects, where the organizational part of the name is certified by the public key of the organization (network) and the itemized part by the public key of the issuer within that organization. The attractiveness of this approach is that it resembles the usage of the existing DNS naming structure. For lookup purposes, it resembles the HANDLE [6] approach, where data might be distributed, but the knowledge of where it is to be fetched from is centralized to an originator location. As long as data is requested from an originator location, freshness and revocation are easy to ensure.

The assumption of connectivity needs to be supplemented by new forms of transport delivery, each carrying its own security twists. End nodes and edge networks might be intermittently connected, known as Delay/Disruption Tolerant Networking (DTN). Regardless of whether the information carried is email or sensor data from disconnected edge devices, the security verification elements need to be part of the stored/cached exchange, and the intermediate networks used need not be trusted or secure. A similar challenge arises from peer to peer swarm distribution, e.g. BitTorrents, where pieces of the information are scattered over a network, and get delivered as result of some dynamic tracking arrangement.

A characteristic of existing internet networks, unlike the closed GSM and 3G operator-controlled networks, is that control information moves on equal basis with the user data, resulting in vulnerability in case of malicious overload (DDOS attacks). The virtualization structures in 4WARD lend themselves to more secure solutions where the critical infrastructure can exchange signaling and dynamic reconfiguration information in isolation from the user traffic proper. Disaster preparation includes the possibility of limiting services and isolating critical subsystems. The rules for storing and deleting information need to be explicitly stated, in particular when it comes to logging transactions.

As we are moving toward an architecture where end nodes often act as servers, for example for P2P networks, the division into control and user planes is not absolute, rather we are talking about a fine-grained trust structure, where some control functions are off-limit and relate to the management of the stratum in question, while other functions are accessible when presenting a valid certificate. The Generic Path concept in 4WARD contains both management and security components, flavors of which are implementation-specific. The central idea is that some resources might come under attack, but network services should still be available. A particular role is held by basic cryptographic functions, which need to be securely available in hardware on each node. While virtual networks generally are independent of the physical networks carrying them and isolated from other virtual networks, it should not be possible to bypass hardware cryptographic primitives with software.

The philosophical aspects of who a user really is cannot be completely ignored. A user will appear in the communication networks in different roles. It is very much a privacy concern that the person hiding behind these roles shall not be connected except when the user explicitly wants so. For making a connection, any eavesdropping man in the middle is questionable—be it legal interception, operator- or business-related datamining, or tampering and content-based intervention. Cryptographic techniques have given rise to a new kind of identity: possession of a private key corresponding to a public key. For privacy reasons, the public keys need not and should not always be the same, instead changing pseudoidentifiers preserve privacy while assuring accountability and non-repudiation. So both users and content can have a verifiable identity, bound to an underlying key-handling arrangement. The common denominator is trust in some form—no technical solution can hide the fact that trust is only weakly transitive, illustrated by key-sharing rings and webs of trust.

The traditional way of securing information has concentrated on containers containing information, and the connections to them. Once the access door is broken down, security is gone. However, a lot more is at stake. The beauty of a computer is that a program is data, and data can be a program. The ugly part is that what looks like content can contain malicious code [19]. From the security angle, allowing data to implicitly contain executable code is a nightmare, and the art of English shell code [20] illustrates how tenuous the difference has become.

An extreme remedy today is the ISP cutting off internet access for a botnet-infected computer, or for a server that is subject to a Denial of Service attack. More importantly, allowing *in principle* the network operator to selectively deliver traffic opens a whole can of worms—possibly the whole security handling should be outsourced to the access network. The principles involved do not match today's reality.

- Cutting off access as a security "solution" displays an act of desperation in face of a design failure

Getting information from "closest" cache is a convoluted question involving both physical capacity and topological information as well as access to privacy/business sensitive location. Secure control over the lookup mechanisms in future networks will be more important than the physical location of data itself.

For commercial purposes, security is calculated risk. Additional features add cost and management, and the average customer does not necessarily require or want to pay for security. Especially privacy is hard to justify in business context, when customers typically like the simplicity of single sign on, oblivious of the personal information that is passed around in the serving networks, sometimes in clear-text form, ready for data-mining.

Another example is the databases of user profiling that search engines like google are able to build up, currently based on cookies planted on the requesting computer. Only technically proficient users clear regularly the cached security-risk implants. These cookies have a legitimate purpose: for the duration of a session, for example a bank wants to know that it is communicating with the same user on the same device.

The 4WARD approach to information handling contains some of the necessary security elements needed. Raising the relative value of metadata describing an information object holds key to one particular aspect overlooked in the existing internet.

For millennia, libraries have been the carriers of knowledge between generations of people and across national boundaries. While search engines crawl the internet today, no Library of Alexandria does the content-based sorting of public information, and any content is probably lost within a decade when the storage address has changed. The ability to verify the authenticity of a piece of information after its certificate has expired is hard, and some attempts toward standardization in this direction have been started, e.g. ETSI PAdES [21].

In conclusion, the 4WARD NetInf approach to information-centric networking has built-in technical approaches for ownership, accountability and freshness of information objects. In reality, a fine-grained technical trust model will have to be matched to a wide scale of user competence.

6.3.4 In-Network Management

4WARD has not specified a model for dynamic naming of virtual slices that appear and disappear, but for monitoring and legal reasons, some meaningful identification method is needed for logging purposes. As a side effect, the end user might have little insight in how he communicates beyond the access network he connects to, and privacy depends on whether end user identification remains at the (insecure?) end node or takes place at the security-capable middleware network element doing authentication. To additionally complicate the picture, the legal requirements on logging traffic are dramatically increasing, e.g. the EU Data Retention Directive 2009 [8], with for example a resulting interesting UK public debate on the "Intercept Modernization Program" and GCHQ "Mastering The Internet", effectively focusing on deep packet inspection and logging of traffic to service providers outside the UK.

The 4WARD architecture framework assumes a lot of exchange of on-line state and available resources between network components. Different operators will have to disclose sensitive information about the state, configuration, and performance of their own networks. This requires a deep change in attitude from existing networks, where only a minimum of information is made available, and where collaboration depends on fairly static service level agreements (SLAs) defining bulk characteristics such as bandwidth. It is possible that future networks consist of 4WARD-like dynamic exchange for SLA contracts on the virtual network level, while SLAs of traditional kind define the bit pipes on the physical level. Crucial elements will be the agreed methods of the Knowledge stratum of monitoring performance, and the actions taken by the Governance stratum when trust has failed.

Network providers and supervisors have strong interest in accurate security log aggregates, as this will allow more precise estimations of the global security situation, in order to take countermeasures and improve operations. There are, however, important privacy concerns, as log data, even in sanitized form, can reveal significant amounts of critical information concerning internal business and network operations.

Fig. 6.1 Example
multidomain network

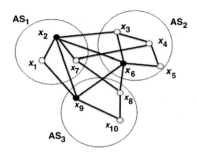

The aggregation protocols used in the context of 4WARD In-Network Management (INM) involve large numbers of network nodes (routers) collaborating to produce performance or security related statistics on huge and generally incomplete
networks. The aggregation process often involves security or business-critical information, which network providers are generally unwilling to share without strong
privacy protection. This appears to be an essential obstacle to enable the INM approach to be deployed in multi-domain, cross-provider settings.

It is therefore important to develop versions of the INM algorithms which are able
to execute without providers private information leaking to outsiders. This is particularly important for network management information, as this generally contains
lots of information about the configuration, operation, load, and performance, of the
providers' internal network. In today's SNMP-based management architecture, protecting this information can be done by point to point encryption and authentication
of the network management traffic. In an INM-based solution this is, however, much
more complicated, since the INM protocols are based on information exchange between large classes of nodes, including nodes that link, e.g., competing provider
domains. Hence, end-to-end encryption and authentication is no longer sufficient
to protect the privacy of internal provider domains, and a different type of solution
must be sought.

One possible approach is to use techniques known from secure multi-party
computation (MPC), where the objective is to compute, in a secure and privacy-
preserving fashion, an arbitrary computable function, distributed among a small
number of fully connected agents. Certain common types of aggregation functions
such as average or sum can be privacy protected without much difficulty. A challenge is to extend this to wider classes such as min/max or threshold functions. The
difficulty is that the INM network calls for very large network graphs, whereas MPC
works for graphs with a small number of nodes only. Hybrid schemes are possible,
however, as indicated by the following example.

Example: In the MPC literature it is known how to compute, e.g., integer comparisons in a fully private manner on a small, fully connected, network [4]. In [14]
we show how to leverage such a construction to compute maximum privately on a
large network, by assuming a small, fully-connected subgraph of "super-nodes".

Consider the multi-domain network in Fig. 6.1.

The network has three AS's, AS_1 to AS_3. Nodes are numbered as indicated. The
super-nodes are the nodes 2, 6 and 9, colored black in the figure. Note that the

subgraph consisting of these three nodes is fully connected. Each node i holds a local state variable x_i. The task is to compute $x = \max(x_1, \ldots, x_{10})$ privately, i.e. in such a way that the only information the nodes learn is that the maximum is x, not which node actually holds the x. This can be done as follows:

On bits, maximum is logical "or", i.e. disjunction. If we know how to compute disjunctions privately then we can compute maximum privately for, say, integers as well, by computing bitwise starting from the most significant bit. Each bit is the disjunction of corresponding bits in a subset of the nodes, namely those nodes that have not yet ruled themselves out from holding the maximum. So, if we can compute disjunctions privately we can compute maximum privately as well, and hence it is enough to think of the x_i as bits.

Unfortunately, computing disjunctions privately is not altogether simple. A trivial solution would be to compute the private sum $y = \sum_{i \in [1,10]} x_i$. Then y is 0 if and only if the disjunction x is 0 as well. But if a node knows y it knows a lot more than intended. For instance it can tell if some other node holds a 1 bit. A solution is to split the sum y in two portions. First compute privately, over the entire network

$$x_{white} = (x_1 + x_3 + x_4 + x_5 + x_7 + x_8 + x_{10} + r_2 + r_6 + r_9) \bmod k$$

as the sum of bits belonging to non-super-nodes plus random seeds r_2, r_6, r_9, generated by the respective super-node. The sum if computed mod k for some sufficiently large prime k. Since y_{white} is computed privately, no information is leaked, at least as long as no collusion of attackers can partition the network. The MPC comparison protocol by Damgård et al. [4] can then be used to test whether or not $x_{white} + (x_2 + x_6 + x_9 - r_2 - r_6 - r_9) \bmod k = x = 0$.

As result we obtain a protocol for computing max which is information-theoretically private against passive, "honest-but-curious" adversaries, under assumptions inherited from the underlying protocols for sum and comparison such as (1) no collusion of adversaries can separate the network graph, and (2) less than half the super-nodes are adversaries.

6.4 Conclusions

Security principles are fundamental requirements, but security itself breaks down into implementation choices and tradeoffs involving usability, business models, management, and fundamental understanding of what is needed. Because the future networks considered bring in paradigm change especially regarding the end to end principle and the role of the infrastructure, a bit of caution is needed. Owners and governments will be as eager as before to exert control over resources and usage. Our current internet security concerns might not be applicable as such when the fundamental architecture has been changed toward a publish–subscribe system, as the 4WARD information centric view implies. On the other hand, finding information will be subject to new security challenges where user privacy might not be highly respected. Trust and availability are granular aspects of security, and this chapter has touched upon these, while leaving relevant questions of interconnection and network neutrality aside.

References

1. D. Atkins, R. Austein, Threat Analysis of the Domain Name System (DNS), RFC 3833 (Informational) (August 2004)
2. Chord, http://pdos.csail.mit.edu/chord/
3. B.-G. Chun, A. Ermolinskiy, K.H. Kim, S. Shenker, T. Koponen, M. Chawla, I. Stoica, A data-oriented (and beyond) network architecture, in *Proc. ACM SIGCOMM*, Kyoto, Japan, August 2007
4. I. Damgård, M. Fitzi, E. Kiltz, J.B. Nielsen, T. Toft, Unconditionally secure constant-rounds multi-party computation for equality, comparison, bits, and exponentiation, in *TCC* ed. by S. Halevi, T. Rabin. Lecture Notes in Computer Science, vol. 3876 (Springer, Berlin, 2006), pp. 285–304
5. J. Day, *Patterns in Network Architecture: A Return to Fundamentals* (Pearson Education, Upper Saddle River, 2008)
6. Digital Object Architecture (DOA): Handle, http://www.handle.net/
7. DNS security extensions, http://www.dnssec.net/
8. EU Directive 2006/24/EC, http://www.ericsson.com/solutions/news/2009/q1/090202-adrs. shtml
9. J. Girão, R.L. Aguiar, A. Sarma, A. Matos, Virtual identity framework for telecom infrastructures. Wirel. Pers. Commun. **45**, 521–543 (2008)
10. Global Environment for Network Innovations, http://www.geni.net/
11. ICANN, Internet Corporation for Assigned Names and Numbers, http://en.wikipedia.org/wiki/ICANN, http://www.icann.org/
12. IESG IAB, IETF Policy on Wiretapping, RFC 2804 (Informational) (May 2000)
13. I. Ioannidis, M. Blaze, J. Feigenbaum, A. Keromytis, The Keynote Trust-Management System Version 2, RFC 2704 (Informational) (September 1999)
14. G. Kreitz, M. Dam, D. Wikström, Practical private information aggregation in large networks, in *Proc. NordSec 2010.* Springer Lectures Notes in Computer Science (in press)
15. M. Larson, D. Massey, R. Arends, R. Austein, S. Rose, Protocol Modifications for the DNS Security Extensions, RFC 4035 (Standards Track) (March 2005)
16. M. Larson, D. Massey, R. Arends, R. Austein, S. Rose, DNS Security Introduction and Requirements, RFC 4033 (Standards Track) (March 2005)
17. M. Larson, D. Massey, R. Arends, R. Austein, S. Rose, Resource Records for the DNS Security Extensions, RFC 4034 (Standards Track) (March 2005)
18. A.A. Levy, H.M. Levy, R. Geambasu, T. Kohno, Vanish: Increasing data privacy with self-destructing data, in *Usenix Security Symposium 2009*, Montreal, Canada, 2009
19. Metasploit—Penetration Testing Resources, http://www.metasploit.com/
20. F. Monrose, G. MacManus, J. Mason, S. Small, English Shellcode, in *ACM CCS09*, Nov 9–13, 2009, Chicago, IL, USA
21. New ETSI standard for EU-compliant electronic signatures, http://www.etsi.org/website/newsandevents/200909_electronicsignature.aspx
22. OAuth, An open protocol to allow secure API authorization, http://oauth.net/
23. Onion routing and Tor, http://en.wikipedia.org/wiki/Onion_routing
24. OpenID, The OpenID Foundation is an international non-profit organization, http://openid.net/
25. PARC, http://mags.acm.org/queue/200901/?pg=8
26. Platform for Privacy Preferences (P3P) Project, http://www.w3.org/P3P/
27. PSIRP, Publish–Subscribe Internet Routing Paradigm, http://psirp.org/publications
28. Rochester Institute of Technology: Floating Cloud Tiered Internet Architecture, see http://www.networkworld.com/news/2010/010410-outlook-vision.html
29. R. Schell, M. Thompson, Platform security: What is lacking (January 2000)
30. J.L. Simmons, Buying You—The Government's Use of Fourth-Parties to Launder Data About "the People", http://www.joshualsimmons.com
31. SPKI Certificate Theory, http://www.ietf.org/rfc/rfc2693.txt, https://wiki.tools.ietf.org/html/rfc2692

32. Stanford Clean Slate, OpenFlow, http://cleanslate.stanford.edu/, http://www.openflowswitch. org/
33. The US Safe Harbor—Fact or Fiction? (2008), http://www.galexia.com/public/research/assets/
34. U.S. European Union Safe Harbor Framework, http://www.export.gov/safeharbor
35. Web ontology language, http://www.w3.org/2001/sw/
36. WHOIS domain search, http://www.ietf.org/rfc/rfc3912.txt
37. P. Zimmermann, Zfone is a new secure VoIP phone software product: Zfone uses a new protocol called ZRTP, http://zfoneproject.com

Chapter 7
Interdomain Concepts and Quality of Service

How We Interconnect Networks and How We Manage Quality of Service (QoS)

Pedro Aranda Gutiérrez and Jorge Carapinha

Abstract One of the key challenges for the Future Internet is the correct defini-tion and implementation of the domain concept. The domain concept is introduced. The interconnection model of the Internet and of current mobile operators is anal-ysed, addressing service ubiquity and interdomain concepts developed in the scope of 4WARD as well. Then, a new interconnection model is introduced, with require-ments, principles, and peering models; architecture elements are addressed, together with interconnection in virtual networks. Special attention is devoted to the still to solve problem of Multidomain Quality of Service, namely The Inter-Provider QoS problem, new challenges and tools for QoS in the Future Internet, and QoS in a network virtualisation environment.

7.1 Introduction

The Internet is partitioned in Autonomous System (ASes) [6], which are inter-connected with each other. This interconnection is governed by peering agree-ments signed between the Internet Service Provider (ISPs) which manage these Au-tonomous System (ASes). Peering agreements include the technical and economical conditions under which the interconnection takes place. Service Level Agreements (SLAs) take a prominent place when negotiating peering agreements. This chapter gives an overview of the current state of affairs in packet network interconnection

P. Aranda Gutiérrez (✉)
Telefonica I+D, Madrid, Spain
e-mail: paag@tid.es

J. Carapinha
PT Inovação, Aveiro, Portugal
e-mail: jorgec@ptinovacao.pt

L.M. Correia et al. (eds.), *Architecture and Design for the Future Internet*,
Signals and Communication Technology,
DOI 10.1007/978-90-481-9346-2_7, © Springer Science+Business Media B.V. 2011

and proposes to extend the current model in order to cope with the complexities expected to arise in the Future Internet. Since QoS is one of the most important components of SLAs, the last section of this chapter is devoted to the evolution of QoS in a Future Internet.

7.2 Domain Concept

What is a domain? Intuitively, the notion of domain in the networking world is tightly linked with partitioning and border. It has also strong associations with business issues, administrative issues, maybe even legal aspects. A domain can be further qualified through a set of one or more properties which the entities of the domain generally have in common. Examples of such are:

- technology (e.g. access network technology)
- protocols (e.g. routing protocol)
- mechanisms (e.g. QoS provisioning and enforcement)
- name and/or address space (e.g. Ethernet, IPv4, IPv6, E.164)
- organisational and business policies (common owner, provider, billing principles, etc.).

These are some of the reasons why a domain has its borders. Although the classification proposed above might imply a hierarchy (in the sense of a 'simple tree structure'), this is not the case. A network element is normally part of many different types of domains. Additionally, it has to be noted that the concept of a domain is recursive in nature, i.e. a domain can consist of other (sub-)domains.

Section 7.3.1 takes a look at the Internet as an example of an environment, where all aforementioned possibilities are represented. The highest macroscopic level of abstraction would be the Autonomous Systems (AS). The AS is a portion of the Internet that is managed using the same set of policies. This definition makes implicit that a domain is administered autonomously based on whatever policies (e.g. routing, resource management, security) are thought appropriate by its respective network administrator, regardless of external conditions and independently of other network domains. The use of ASes is a key enabler of scalability of Internet routing on a global level.

7.3 Interconnection Models

In order to study the interconnection of domains, both in today's networks and in the Future Internet, the domain concept needed to be defined. This section studies the interconnection models of the Internet and their evolution both in the Mobile World and from the point of view of Service Oriented Architectures. All the concepts provided by the different worlds converge in the interdomain models proposed by 4WARD.

Fig. 7.1 Peering types through Internet Exchange Point (IXP) and using private peering

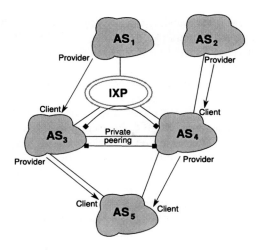

7.3.1 Interconnection in the Internet

The Internet is partitioned into domains known as ASes [6]. These domains are interconnected with each other. From a technical point of view, ASes exchange routing information using the Border Gateway Protocol (BGP-4) [12]. Looking at the problem from a political or business point of view, the interconnection of Internet domains is governed by peering agreements signed between ISPs. Peering agreements define the technical and economical conditions under which the interconnection takes place.

ISPs have established a ranking among themselves, which somehow reflects their importance by the proximity to the core of the Internet, which is defined as a fully-meshed core layer of so-called **Tier 1** providers. Below these, different levels of providers are partially interconnected and provide ultimately Internet access to end users. While *Tier 1* providers exchange full routing information, others may establish a more complex net of relationships with the ASes they are connected with. As shown in Fig. 7.1, there are two basic relationships between ASes:

1. **Client–Provider**
2. **Siblings** (or **Peer-to-peer**, although this definition is less used because of the confusion introduced by so-called *P2P file exchange networks*).

These relationships between ASes reflect their commercial relationships and how they exchange traffic among them. Siblings exchange routing information about themselves and their clients. In the extreme case of Tier 1 ASes, this implies exchanging the full routing tables. Non-Tier 1 ASes usually follow the following set of rules:

1. Send all the clients' traffic to the clients
2. Send all traffic directed to siblings and their clients to siblings
3. Send the rest of the traffic to providers.

These rules translate into routing policies that have to be programmed into Internet Core Routers [4] in order to control the BGP-4 information exchange. This interference into the protocol's natural behaviour has severe side effects on the its convergence characteristics [1, 2].

At the beginning, ASes were interconnected using point-to-point lines which evolved to bundles of point-to-point lines as the traffic grew. Eventually these bundles were substituted by point-to-point lines with higher capacity and when these carried an amount of traffic above certain thresholds, bundles were introduced again, and so on. This evolution was welcome by long distance carriers, because these lines meant a growing income, but not by ISPs. The cost of interconnection not only includes the transmission costs. An essential component of peering agreements between ISPs is the charging procedure for data exchanged among them.

The Internet Exchange Point (IXP) concept was born as a means for ISPs to cut down interconnection costs. Instead of long distance lines, ISPs would establish shorter connections to a common infrastructure. And instead of paying to upstream providers, ISPs could use IXPs to exchange traffic on a fair basis. Despite this, depending on the by-laws governing the operation of a given IXP and the cost of connecting to it, ISPs also resort to private peering agreements.

7.3.2 Interconnection in the Mobile Data World

As mobile data services became popular and mobile terminals started to gain market share, the GSM Association (GSMA) recognised the need to provide an efficient way to support data services for roaming users at reasonable costs. Two models emerged for the Interconnection in the Mobile Data World: the GSM Roaming Exchange (GRX) and the IP packet eXchange (IPX). Both are defined in the GSMA recommendation IR.34 [3]. The GRX evolved from the Internet Interconnection Model in order to provide data services for roaming mobile users. GRXs interconnect mobile operator IP backbones for Global System for Mobile Communications (GSM) roaming. They provide a centralised IP-routing network for interconnecting GPRS and UMTS networks. It provides a *pure IP infrastructure*, where client border gateways connect the different operator backbones to a GRX backbone on the IP level. This basic model can be seen as a parallel Internet for Mobile Users. The GRX provides the basic functionality on an IP Best Effort (BE) infrastructure. IR.34 also defines with the IPX a more complex network infrastructure that is needed to implement QoSs based IP services and

1. Hubbing services for Multimedia Messaging System (MMS)
2. SMS Sigtran interface for Signalling
3. VoIP/SIP-I.

The challenge of the GRX model is to provide a consistent service level, in terms of QoS and pricing, in order to leverage user experience when using the home network and when visiting foreign networks. Compared with pure IP interconnection and with the GRX model, the IPX model provides an infrastructure implementing service interoperation.

7.3.3 Service Ubiquity

Service Ubiquity is the main objective of the IPSphere Model [9]. IPSphere provides a framework for service interworking including all aspects of the value creation chain. The main focus is to dissolve barriers to consumption of services enabled by New Generation Networks (NGNs). To achieve this, it exposes network capabilities in a business relevant manner that incorporates business parameters alongside technical parameters in published offers. IPSphere enables network services to be offered and consumed in the same manner as IT services are. It heavily relies on Service Oriented Architecture (SOA) and defines Web Services which incorporate *federation* mechanisms to ensure pan-operator ubiquity. IPSphere also provides for service differentiation according to performance and trust levels. IPSphere requires a demanding infrastructure, but this is necessary due to the complexity of the interactions between parties.

The IPSphere Model reaches beyond the Internet interconnection model, which basically consists in exchanging routing information on an infrastructure with a common network layer. It points towards defining interworking interfaces at higher layers in order to allow applications in different domains to communicate and construct new, richer applications for the end user. This direction will be necessary in order to provide an infrastructure capable for Future Internet environments.

7.4 Towards a New Interconnection Model

All interconnection models described above have one thing in common. They can be considered horizontal interconnections in the sense that they define the interconnection of equivalent infrastructures at equivalent layers in the protocol stack. The only hierarchical relationship which might help identify a vertical interconnection between two networks or network providers is the Provider–Customer relationship in the IP Interconnection. A vertical dimension in the sense of different layers in the protocol stack is not considered currently, but is urgently needed when collapsed protocol stack networks promised by Generalised Multi-Protocol Label Switching (GMPLS) [8] become commonplace.

Taking into account that advanced features like radical virtualisation, protocol stacks as plug-ins and highly heterogeneous networks are going to be an essential component of the Future Internet, it becomes obvious that an Interconnection Model is needed that is more flexible compared with the Interconnection model of the current Internet. The requirements for this model are explored in this section. They are presented using proposals and developments from 4WARD.

7.4.1 Interconnection Requirements

In a world, which aims at being multilayer and multitechnology, with full interworking between heterogeneous domains, horizontal peering models like those deployed in today's networks are not going to be able to fulfil all requirements. Even today, many services would require cross-layer interaction. An example for this are multimedia services in general and IPTV services in particular. They are either provided in logically and physically independent networks, which are deployed in parallel to the Internet infrastructure. Although not completely infeasible, this kind of services cannot be deployed over the current Internet because consistent QoS support is lacking there. Another example for the infeasibility of a common infrastructure are the GRX and IPX efforts by the GSM Association, which were briefly explained in Sect. 7.3.2.

7.4.1.1 Interworking Principles

In an environment, where the interworking of heterogeneous networks is key, besides specific Architecture Elements to implement the interworking functionality per se, basic interworking principles have to be defined in order to assure an orderly and efficient design and deployment process. In the 4WARD context [17], where the network environment is heterogeneous in different ways, interoperability assures that the means to overcome this heterogeneity are in place, assuring that applications will run satisfactorily end-to-end. Heterogeneity can be at many different places and levels of networking. The scope and the focus of interoperability can potentially be adapted to different networks, and the issue of achieving interoperability between those networks is a matter of providing a suitable interface at the border between them. In the scope of the Future Internet, networks might not only have a peering and/or a transit interprovider relation as presented in previous sections, but also a user-provider relation where 'layers' of networking (or rather functionality) are stacked on top of each other using a common physical infrastructure.

In order to support interoperability and enable that applications can operate end-to-end with performance guarantees, 4WARD has identified following Interoperability Principles

1. **Interoperability Principle #1:** Application characteristics shall be preserved at the border between networks.
   In the Future Internet, a wide set of applications and services can be foreseen, each one with its own requirements (network performance parameters, security constraints, etc.). As it is assumed that each network and transit network provides and implements the necessary properties, functions, and technologies for proper application support, the domain border must not compromise the application characteristics below a level which is deemed unsatisfactory by the users of an application in order to preserve the end-to-end application characteristics.
2. **Interoperability Principle #2:** The network border should only need to preserve characteristics for those specific applications that run across that network border.
   In order to assure the preservation of the application characteristics in the end-to-end path, the networks will agree in advance the characteristics that should be preserved for each application (and the costs) and will configure the domain or network borders accordingly during this negotiation phase. In order to avoid

that each domain border must need to give support for any possible application, which would lead to a maximised and one-size-fits-all solution, the domain border should be allowed to be tailor-made depending on what type of applications that are in use across that border. Therefore the borders will only guarantee the characteristics for the applications they know in advance. So, if new application types must be preserved, new negotiations should be carried out. This does not mean that other applications are filtered or discriminated; it just means that they are not preserved according to specific targets since these targets are not negotiated.

3. **Interoperability Principle #3:** Application characteristics shall be generically, explicitly and commonly encoded at a network border.

   The diverse set of applications characteristics needs to be modelled into a generic set of characteristics to avoid direct application dependencies, as well as scalability and complexity issues that would arise over time when new applications will appear. Examples of such generic characteristics are end-to-end delay, jitter, privacy, and packet loss/error rate. Therefore, the applications should be grouped into application types or classes of services (e.g. class of services for real-time and streaming applications instead of Skype, Messenger, YouTube or specific operator applications). Explicit, generic and common encoding avoids dependencies on specific solutions, technologies and standards. It also avoids translation mechanisms being mandated at the domain, which would generally not scale and which could lead to performance bottlenecks. Please note that nothing prohibits that the same encoding can be used within the networks and the functions and technologies being deployed in those networks involved in the end-to-end path. The end users will also benefit from this principle, since they will not need to adapt their application specifications to the network they are using since the description will be commonly understood by the domains that provide guarantees for such application type. This allows the end user to enjoy their guaranteed performance application characteristics whatever the network they use without the need to change the application description which is especially important in future environments with high mobility degree.

4. **Interoperability Principle #4:** Each network border should provide the necessary capabilities and means to compose the networks being connected via the domain border.

   In order to preserve the application characteristics across the end-to-end path, the composition between different networks will be required. Composability is an aspect of both dynamicity and self-management. Generally, networks shall be able to compose across a domain border, i.e. to dynamically establish a satisfactory level of trust, and to negotiate an SLA (Service Level Agreements) regulating how resources and services in each of the participating networks can be used across the domain border (and which may include how compensation shall be done, as well as covering other aspects of security, management, etc.). To what extent the trust and the SLAs are dynamically established can however be determined case-by-case. A minimum level of composability is for further study.

7.4.2 New Peering Models

In order to overcome the current architectural limitations, new interconnection models are needed. The current model foresees two main interconnection modes: horizontal and vertical peering. These terms refer both to business relationships and hierarchical positions in the Internet. In this line of thought, sibling ASes establish a horizontal peering to exchange traffic, whereas a client-provider relationship at the AS level is a vertical peering. However, when a scenario of interconnected and interoperating heterogeneous networks, like in 4WARD's vision of the Future Internet, this model needs to be expanded for the sake of flexibility.

7.4.2.1 Architecture Elements

Figure 7.2 shows two 4WARD domains including their stratum model as defined in [16].[1] Domains are composed of different strata, which are controlled by a common Governance and Knowledge Stratum. Strata are sets of logical Nodes that are connected through a Medium and encapsulate functions that are distributed over the nodes. These functions are provided to other strata through two well known interfaces:

- The Service Stratum Point (SSP) provides the services to the other strata located on top of or below the stratum and within a domain.
- The Service Gateway Point (SGP) offers peering relations to other strata of the same type.

Al illustrated in Fig. 7.2, strata communicate with the outside world via the SGP. Within a domain, strata communicate using the SSP. With this definition in mind, three peering types can be defined:

1. **Vertical peering via the SSP**, which provides interoperation between strata within a domain.
2. **Horizontal peering via the SGP**, which provides interconnection between domains, when interconnecting functionally equivalent strata (as expressed by the $SGP_x \leftrightarrow SGP_x$ in Fig. 7.2).
3. **Transversal peering**, which implements the interconnection between functionally different strata of different domains (as expressed by the $SGP_x \leftrightarrow SGP_y$ communication in Fig. 7.2).

When the stratum is implemented using Netlets, interoperability is implemented in Interop netlets [16], which are specialised in connecting different types of networks. Interop Netlets are shown in Fig. 7.3 in conjunction with regular and control netlets, in order to highlight their complexity. This complexity is needed in order to mediate between similar and/or different network architectures and as such have to

---

[1] See Chap. 4 for a detailed description of the 4WARD.

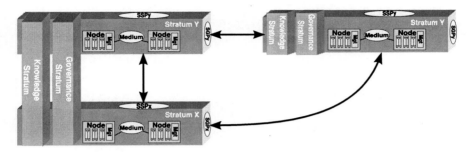

Fig. 7.2 Peering in the stratum model

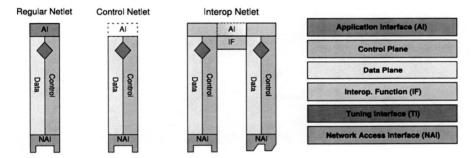

Fig. 7.3 Netlet types: components needed to provide interoperation at the Netlet level

be prepared to implement:

1. address mapping
2. protocol translation
3. content transcoding

and other interoperability functions.

7.4.2.2 Interconnection in the Virtual Networking World

Virtualisation is going to play a key role in the Future Internet and means are needed to interconnect technologically heterogeneous Virtual Network (VNets) at different levels. These different levels are needed, because the different hierarchical levels in the network are coupled with different provider roles:

- the *Infrastructure Provider* (InP) providing the hardware infrastructure
- the *VNet Provider* (VNP) providing VNets on top of this hardware infrastructure
- the *VNet Operator* (VNO) providing the actual VNets to the end user.

Fig. 7.4 The Folding Link concept

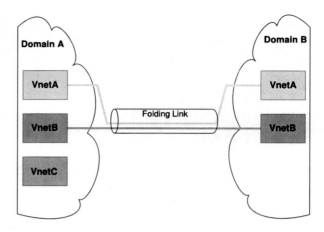

Fig. 7.5 The Folding Node concept

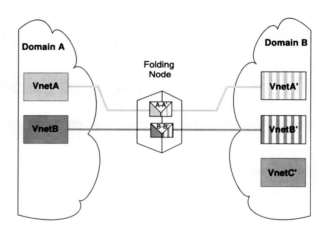

The 4WARD project has foreseen two different kinds of interconnection points in the VNet ecosystem [18]. Folding Nodes (FNs), as shown in Fig. 7.5 will interconnect heterogeneous infrastructures. They provide interworking functions between them. Folding Links (FLs) are used when the main objective is to provide connectivity over administrative boundaries of otherwise homogeneous infrastructures, see Fig. 7.4.

Horizontal and transversal peering take place at these points. Folding Nodes may provide interworking functionalities between dissimilar VNets and thus implement the $SGP_x \leftrightarrow SGP_y$ interface functionality. Folding Links connect nodes with an equal set of VNets, which can be modelled as a $SGP_x \leftrightarrow SGP_x$ interface functionality.

7.5 Inter-domain QoS

7.5.1 Introduction

There is little doubt that *Quality of Service* (QoS), either as a measure of the quality experienced by the user or as a set of tools to ensure predictable performance, will constitute a key requirement for the Future Internet. Pervasiveness of network-based applications and emerging trends, such as cloud computing, relying on predictable network performance, will contribute to exacerbate the need for QoS.

Obviously, QoS issues, and the respective solutions, will not remain unchanged. QoS has always been and will likely continue to be a moving target, depending on the services and applications, as well as characteristics of the underlying network infrastructure. New ways of dealing with QoS will be enabled by novel networking concepts and techniques. New challenges and requirements will certainly emerge.

For a number of years, QoS represented one of the most active areas in networking research. The main outcome of this effort has been a toolkit, or a set of building blocks, that network providers can combine and use to accomplish specific service requirements in single network domains [13].

A universal end-to-end framework to provide QoS to large scale heterogeneous networks has not been materialised to date and it is doubtful that it will ever be. This largely results from the decentralised nature of the Internet—because no single entity is responsible for administering the Internet globally, the enforcement of a common policy for resource management across multiple heterogeneous network domains is complicated, if possible at all. Unfortunately, from the end user perspective, the experienced quality of service will always be end-to-end and depend on the cumulative effect of the quality impairments across every network domain from source to destination.[2] It has also become clear that QoS alone does not necessarily reflect the service quality perceived by users, thus an additional concept is needed to properly describe the user perspective—that concept is usually called Quality of Experience (QoE) and is supposed to provide a subjective measure of the user experience. Compared to QoS, QoE has a broader scope as it takes into account every factor that potentially contributes to user satisfaction and ultimately determines user's perception of the service value such as flexibility, mobility, security or cost.

QoS in inter-domain scenarios remains a challenge today and Future Internet will likely bring additional challenges. The next sub-section discusses the inter-domain QoS problem in general, whereas the subsequent sub-sections are focused on Future Internet scenarios and evaluate the possible impacts of network virtualisation on inter-domain QoS.

---

[2]ITU-T Recommendation Y.1541 defines QoS target values in terms of delay, delay variation, loss and error rate, for 5 different QoS classes.

Fig. 7.6 QoS in a
multi-domain scenario

7.5.2 The Inter-provider QoS Problem

In the general case, an end-to-end path from source to destination will traverse multiple network administrative domains, managed by different independent (often competitor) operators.

In each network domain, QoS will be ultimately determined by how network resources are managed and allocated to each traffic flow. In a single administrative domain, the respective network operator is supposed to define and enforce the adequate resource management policies, in order to achieve specific QoS goals. Certain QoS guarantees can be provided to specific classes of traffic by appropriately combining resource dimensioning, traffic admission control and traffic class differentiation.

Controlling end-to-end QoS in a multi-domain scenario is usually much more complicated. Figure 7.6 illustrates the general case, in which the path between two end points crosses n domains. The overall end-to-end target delay, delay variation, loss probability and throughput values are denoted as D, J, P, T, respectively, whereas the corresponding values in QoS domain i, as d_i, j_i, p_i, t_i. For any individual packet, in order to accomplish the end-to-end target values, the following conditions must be fulfilled:

- $\sum d_i \leq D$
- $\sum j_i \leq J$
- $\pi(1 - p_i) \geq 1 - P$
- $t_i \geq T$

To support QoS meaningfully across multiple domains, it is essential that QoS metrics are defined in a consistent manner. In practice, Service Level Agreements (SLA) are usually based on statistical information measuring the average network behaviour over a sufficiently long period of time (e.g., 99.9% of the packets are delayed no more than 20 ms over a period of 10 minutes).

Several proposals have been put forward to deploy end-to-end QoS control in a multi-domain scenario:

The FP5 Mescal Project [7] proposed solutions for the deployment and delivery of inter-domain QoS. MESCAL is focused on the business relationships, based on Service Level Specifications, between customers and IP Network Providers, and between IP Network Providers, to enable QoS-based IP connectivity services. Each Network Provider establishes agreements with direct neighbours to allow the extension of QoS guarantees over multiple domains. A hop-by-hop cascaded model is adopted for interactions between providers at both service and network layers. The interdomain routing protocol, BGP-4, in a QoS-enhanced variant, can be used to support dynamic inter-domain Traffic Engineering. Thus, QoS can be provided from source to destinations that may be several domains away.

The FP6 EuQoS project proposes a solution based on the definition of a virtual network layer, which decouples network decisions from network technologies [11]. In each network domain, a resource manager is in charge of managing QoS, coordinating admission control decisions, managing peering agreements with neighbouring domains and controlling the inter-domain routing process. EQ-BGP is the inter-domain protocol proposed to select and advertise the paths for the different classes of service. EQ-BGP extends standard BGP-4, among other things, by including an optional path attribute, QOS_NLRI, that conveys information about the path QoS characteristics.

A potential problem with the above proposals is the fact that they are based on the extension of BGP to carry QoS-specific information, which raises concerns in terms of convergence and scaling. Another problem is the dependence from a "common language", by means of which a universal set of classes of service would be defined and used in all network domains. Given the heterogeneity of technologies and architectures, and mainly the different policies for resource management across different network provider domains, this would be a very difficult, and probably unrealistic, goal to achieve.

In addition to technical hurdles, the lack of sound business models, as well as feasible QoS accounting and charging approaches have contributed to disincentive service providers from deploying inter-domain QoS solutions [10]. In fact, QoS is not a purely technical problem and therefore cannot be tackled exclusively by technical means. The definition of adequate business models is an essential requirement to successfully deploy QoS. This is particularly true in inter-domain scenarios: without an attractive business model, the incentives for the various players to fulfil QoS requirements will not be clear, thus probably inter-domain end-to-end QoS will never materialise.

7.5.3 QoS in the Future Internet—New Challenges and Tools

QoS mechanisms and tools should be revisited in the light of emerging networking concepts. Future Internet is expected to support a myriad of applications and services, with a very wide range of characteristics and requirements. Deterministic performance is often included in the Future Internet "wish list".

It is a widely held belief that over-provisioning network capacity alone will be sufficient to guarantee QoS in most situations, thus circumventing the need to use complex and costly QoS mechanisms. However, over-abundance of network resources cannot be taken for granted either for economic reasons (e.g. massive investments required by widespread deployment of fibre in access networks), or for inherent technological limitations (e.g. spectrum availability in wireless networks). Also, even in the cases where over-provisioning is a feasible solution, the likelihood of congestion at some point in the network is not eliminated, and may actually occur as a result of several possible causes—malicious combined effect of multiple traffic sources (e.g. Distributed Denial of Service (DDoS) attacks), traffic shifts within the network due to popular contents, redirection of traffic caused by network failure, or even fluctuations of traffic volume on a link as a result of normal stochastic behaviour. On the other hand, it is a historically verifiable fact that growth of network capacity tends to be accompanied by a corresponding increase in users bandwidth demand. At least 50% annual increase of capacities and traffic has been observed over more than a decade in fixed and wireless network platform [5].

As resource over-provisioning, per se, is not the answer to the QoS problem, it is clear that some form of QoS control will be required in future networks. Unfortunately, for network operators, implementation of QoS techniques on a large scale is far from trivial and is often seen as a cost, which many network operators would prefer to avoid, or at least minimise. Many network administrators choose not to enable QoS functions in their networks because they are difficult to configure properly, requiring a full understanding of the complex mechanisms behind and the dynamics of the traffic.

In this scenario, In-Network Management (INM) constitutes a promising tool to simplify and improve scalability of QoS management operations. INM overcomes traditional network management limitations by means of real-time monitoring functions and automated configuration management. INM can support adaptation and self-configuration of large-scale networks according to external events and permits low-cost operation [14]. These capabilities can be used to simplify QoS management by pushing QoS-related management capabilities into the network and distributing QoS management logic across all nodes. In INM, the architecture of each network node contains a dedicated QoS management Functional Component (QoS dmFC), which handles all functions related to guaranteeing a specific level of quality of services offered to the users, based on three groups of management capabilities: accessing the physical resources, cross-layering for QoS parameters and/or data, composite metric calculation [15].

As discussed before, user's satisfaction cannot be measured solely by QoS, as it requires a broader set of metrics, which are usually combined in the QoE concept. In this regard, NetInf can bring important benefits in terms of upgrading quality of experience perceived by the user [19]:

- Efficient large-scale distribution: NetInf provides an application-independent service for distributing information on a large scale that is robust to rapidly changing demand.
- Increased information availability: since requests for information in NetInf can be served by any host holding that information, there is no dependency on connectivity to a particular server or set of servers. Also, availability of information is improved since information identifiers are independent of their location, which means that the information can be relocated.
- Increased security: mechanisms for author authentication and origin verification, as well as mechanisms for checking content integrity, are integrated into the network service (contrary to today's network where security is mostly based on trusting the server delivering the information).

Finally, network virtualisation, addressed in the following sub-section, offers a layer of abstraction that hides the specificities of the network infrastructure and

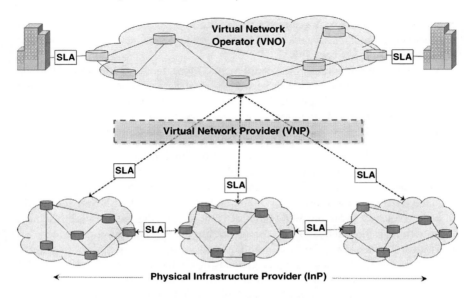

Fig. 7.7   Single virtual network domain, multiple infrastructure domains

enables a unified view of network resources spread over multiple domains, thus overcoming the traditional limitations of QoS control in this kind of scenarios.

In summary, it is clear that the capability to control QoS will remain a fundamental requirement for the networks of the future. New networking concepts and techniques will enable new ways of dealing with QoS and offer new tools to improve QoE. However, several issues must be reassessed in the light of emerging networking concepts and tools—how network resources can be dynamically controlled and shared, how end-to-end QoS can be provided over heterogeneous networks or across multiple provider domains, how to guarantee scalability, how QoS can seamlessly coexist with features like mobility and security.

7.5.4 QoS in a Network Virtualisation Environment

Network virtualisation will likely constitute a key component of the Future Internet. As described before, many network domains will be composed by virtual, rather than physical, network resources. This fact will have a significant impact on how QoS can be provisioned and controlled, both in intra-domain and inter-domain environments.

The basic network virtualisation environment is represented in Fig. 7.7. The roles played by each of the actors has already been discussed before and will not be repeated here. In this scenario, several types of business relationships, based on an SLA specified by QoS parameters, can be established: between the end user and the VNO, between the VNO and the InP (for the sake of simplicity, a direct business relationship, i.e. an SLA, is assumed between the VNO and the InP during the virtual network operational phase) and, in the case where the virtual network spans multiple physical domains, between InPs.

Virtualisation raises new QoS requirements. More than QoS differentiation, virtualisation requires QoS isolation between virtual networks. As a matter of fact, a virtual network should replicate the behaviour of a network supported by physical resources, in all respects, including deterministic performance. Thus, QoS has to be handled at two different levels in a virtualisation-based environment:

- Inside a virtual network, any QoS policy can in principle be autonomously defined and deployed by the VNO (for example, classes of service differentiation for voice and data). Intra-VNet QoS mechanisms must not be constrained by the characteristics of the substrate or by the traffic running in other VNets sharing the same resources. From the VNO point of view, the fact that the network resources are virtual, rather than physical, should not impose any specific limitation.
- At the substrate level, QoS isolation between virtual networks is a key requirement, as any QoS policies and mechanisms inside the VNet itself cannot be built without adequate guarantees of availability of resources. This means that the full amount of capacity subscribed by the VNO should be permanently available, regardless of the traffic originated by concurrent virtual networks sharing the same physical resources. A customer contracting a network service from a VNO should expect a performance level similar to what would be obtained with exclusive physical resources.

This 2-level problem complicates inter-domain QoS. While many of the QoS-related problems posed by network virtualisation can be solved by applying classical QoS approaches, the added complexity and possible impact on scalability are issues that deserve further analysis and evaluation. Even in the cases where a single VNO provides the end-to-end path between customer premises, the underlying substrate path will probably be based on multiple disparate physical network domains, which makes the identification of responsibility for an SLA violation, among the multiple players involved in delivering the end-to-end service, particularly challenging.

On a different perspective, network virtualisation may facilitate the management of QoS by decoupling virtual networks from the underlying network infrastructure. This is particularly relevant if multiple physical infrastructure domains are involved, in which case the establishment of a consistent QoS policy across multiple heterogeneous networks would traditionally represent a complicated challenge. By providing a level of abstraction that hides the physical characteristics of the infrastructure, network virtualisation enables a smooth and simplified management of resources for the VNO. In addition, the possibility offered by network virtualisation to allocate or deallocate virtual resources (e.g. link bandwidth), or modify the network topology on demand, represents a powerful tool to dynamically control network resources and ultimately QoS.

In summary, network virtualisation is a promising tool to handle QoS in scenarios based on network resources located in multiple infrastructure domains, thanks to the abstraction layer that decouples virtual from physical resources. However, the isolation of virtual networks and fine grained control of resources are requirements that raise scalability issues and deserve further study.

7.6 Conclusion

Interconnection is one of the corner stones in the success of the current Internet. Despite limitations, it has provided the tools for a sustainable growth in the last 30 years. The paradigm shift implied by the Future Internet has a great impact on the way interconnection has to be tackled. Virtualisation will exacerbate one of the key issues in providing glitch-free Interconnection: traceability for the purpose of commercially sound Interconnection and Service Level Agreements between providers. The provision of multi-domain Quality of Service is one of the new frontiers in Telecommunications in this sense.

In order to make all these challenges feasible, 4WARD has identified a set of principles which will help overcome many of the growth problems currently observed in the Internet. It has also provided the first conceptual building blocks for interdomain interconnection, both from a purely architectural point of view and applied to the interconnection of virtualised networking environments.

However, the most significant change in the design philosophy of the Future Internet is that Interdomain Issues are being considered as an integral part of its Design and not as an add-on *a posteriori*.

References

1. T.G. Griffin, G. Wilfong, An analysis of BGP convergence properties, in *Proc. of SIG-COMM'99* (ACM, New York, 1999), pp. 277–288
2. T.G. Griffin, F.B. Shepherd, G. Wilfong, The stable paths problem and interdomain routing, IEEE/ACM Trans. Netw. **10**, 232–243 (2002)
3. GSM Association, Inter-Service Provider IP Backbone Guidelines, IR.34 (2008), http://www.gsmworld.com/documents/ir3444.pdf
4. S. Halabi, *Internet Routing Architectures*, 2nd edn. (Cisco Press, Indianapolis, 2000)
5. F. Hartleb, G. Halinger, S. Kempken, Network planning and dimensioning for broadband access to the Internet regarding quality of service demands, in *Handbook of Research on Telecommunications Planning and Management for Business* (IGI Global Publishing, Hershey, 2009), pp. 417–430
6. J. Hawkinson, T. Bates, Guidelines for creation, selection, and registration of an Autonomous System (AS) (1996), http://tools.ietf.org/html/rfc1930
7. M.P. Howarth, P. Flegkas, G. Pavlou, N. Wang, P. Trimintzios, D. Griffin, J. Griem, M. Boucadair, P. Morand, A. Asgari, P. Georgatsos, Provisioning for interdomain quality of service: The mescal approach, IEEE Commun. Mag. (2005)
8. IETF Multiprotocol Label Switching (MPLS) Working Group, http://www.ietf.org/dyn/wg/charter/mpls-charter.html. Accessed 14 Feb 2010
9. IPsphere. http://www.tmforum.org/ipsphere/. Accessed 14 Jun 2010
10. C. Macian, L. Burgstahler, W. Payer, S. Junghans, C. Hauser, J. Jaehnert, Beyond technology: The missing pieces for QoS success (2003), http://www.ikr.uni-stuttgart.de/Content/Publications/Archive/cm_RIPQoS03_SIGCOMM_36310.pdf

11. X. Masip-Bruin, M. Yannuzzi, R. Serral-Gracià, J. Domingo-Pascual, J. Enríquez-Gabeiras, M. Ángeles Callejo, M. Diaz, F. Racaru, G. Stea, E. Mingozzi, A. Beben, W. Burakowski, E. Monteiro, L. Cordeiro, The EuQoS system: A solution for QoS routing in heterogeneous networks, IEEE Commun. Mag. (2007)
12. Y. Rekhter, T. Li, S. Hares, A Border Gateway Protocol 4 (BGP-4) (2006), http://tools.ietf.org/html/rfc4271
13. J. Soldatos, E. Vayias, G. Kormentzas, On the building blocks of quality of service in heterogeneous IP networks, IEEE Commun. Surv. Tutor. (First Quarter 2005)
14. The 4WARD Consortium, D4.1 definition of scenarios and use cases (2008), http://www.4ward-project.eu/
15. The 4WARD Consortium, D4.2 in-network management concept (2009), http://www.4ward-project.eu/
16. The 4WARD Consortium, Deliverable D-2.2 "Technical Requirements" (2009), http://www.4ward-project.eu/
17. The 4WARD Consortium, Deliverable D-2.3.0 "Architectural Framework: New Release and First Evaluation Results" (2009), http://www.4ward-project.eu/
18. The 4WARD Consortium, Deliverable D-3.1 "Virtualisation Approach: Concept" (2009), http://www.4ward-project.eu/
19. The 4WARD Consortium, Second NetInf architecture description (2010), http://www.4ward-project.eu/

Chapter 8
Managing Networks

Daniel Gillblad and Alberto Gonzalez Prieto

Abstract We propose a solution for management, In-Network Management, which is based on decentralization, self-organization, and autonomy of management processes. Its key idea is that management stations outside the network delegate management tasks to the network itself, supporting large-scale networks that self-configure, dynamically adapt to external events and allow for low-cost operation. We discuss challenges, benefits, and approaches to In-Network Management. We present an architectural framework suitable for different levels of embedding within the network elements. Examples of novel algorithms supporting real-time monitoring in a distributed manner are presented, and self-adaptation schemes for resource control are discussed; we outline real-time monitoring of network-wide metrics, group size estimation, data search, and anomaly detection. We conclude that robust, distributed algorithms can be devised for a multitude of management tasks without introducing excessive amounts of overhead in the networked devices.

8.1 Introduction

To cope with the challenges involved in maintaining ever larger, heterogeneous networked systems, the 4WARD project has worked toward novel management instruments to operate the future Internet. We have defined

1. A new framework to support distributed management operations.
2. A set of distributed management algorithms for network monitoring and self-adaptation.

D. Gillblad (✉)
SICS—Swedish Institute of Computer Science, Stockholm, Sweden

A. Gonzalez Prieto
KTH—Royal Institute of Technology, Stockholm, Sweden

L.M. Correia et al. (eds.), *Architecture and Design for the Future Internet*,   151
Signals and Communication Technology,
DOI 10.1007/978-90-481-9346-2_8, © Springer Science+Business Media B.V. 2011

Together, these building blocks provide a foundation for performing network management autonomously and distributed within the network.

8.1.1 Limitations of Existing Approaches

Traditional management solutions typically execute in management stations that are outside the network. These stations interact with the managed devices, mostly on a per-device basis, in order to execute management tasks. During network operation, for instance, a management station periodically polls individual devices in its domain for the values of local variables, such as counters or performance parameters. These variables are then analyzed on the management station by management applications to determine the set of actions to execute on each of the managed devices.

This paradigm of interaction between the management system and the managed network has been used by traditional management frameworks and protocols, including SNMP, TMN and OSI-SM. The paradigm has proved successful for networks of moderate size, whose states evolve slowly. In particular, their configuration rarely needs to be changed and no rapid interventions are required. However, many of today's emerging networks depart from these characteristics. We envision that in the future Internet, particularly with its wireless and mobile extensions and approaching an Internet of things, networks will include millions of network elements, whose state will be highly dynamic and whose configuration will need to adapt on a continuous basis.

We have primarily addressed the aspects of the traditional management paradigm that lead to poor scaling, inherently slow reactions to changing network conditions, and the need for intensive human supervision and frequent intervention.

8.1.2 The INM Approach

Today's deployed management solutions have two main characteristics. First, managed devices generally present simple and low-level interfaces. Second, management stations interact directly with managed devices typically on a per-device basis. There is no interaction among managed devices for management purposes and those devices have little autonomy in making management decisions. As a consequence, managed devices are "dumb" from a management standpoint.

Historically, management solutions where designed with these two characteristics in order to have network elements of low complexity and to achieve a clear separation between the managed system that provides a service and the management that performs configuration and supervision. This also allows for simple, hierarchical structuring of management systems.

We propose a clean slate solution to network management. By this we mean a way of envisioning and engineering management concepts and capabilities that abandons the two characteristics described above.

The approach we propose is called *In-Network Management* (INM). Its basic enabling concepts are decentralization, self-organization, and autonomy. The idea is that management tasks are delegated from management stations outside the network to the network itself. The INM approach therefore involves embedding management intelligence in the network, or, in other words, making the network more intelligent. The managed system now executes management functions on its own. It performs reconfiguration and self-healing in an autonomic manner.

In order to realize this vision, management functionality is associated with each network element or device, which, in addition to monitoring and configuring local parameters, communicates with peer entities in its proximity. The collection of these entities creates a management plane, a thin layer of management functionality inside the network that performs monitoring and control tasks.

In terms of the traditional FCAPS (fault, configuration, accounting, performance, and security) telecommunication network management model, we have concentrated on performing fault, performance, and configuration management tasks within the network itself. The solutions we have developed can also be used to support accounting and security, e.g. through providing key information for billing.

The potential benefits of the INM paradigm include the following properties:

- A *high level of scalability* of management systems, e.g. in terms of short execution times and low traffic overhead in large-scale systems. This will allow for effective management of large networks.
- *Fast reaction times* in response to faults, configuration changes, load changes, etc. This increases the adaptability of the network. This will lead to a high level of robustness of the managed system.
- A *high business value* for INM technologies through reducing both capital and operational expenditures.

A possible drawback of this paradigm is that processing resources for management purposes must be available in the network elements, potentially increasing cost and network element complexity. However, the reduced operational costs of such a system are likely to offset this increased initial capital expenditure.

8.1.3 Scope and Contributions

One can consider a network management system as executing a closed-loop control cycle, whereby the network state is estimated on a continuous basis (i.e., situation awareness), and, based on this estimation, a process dynamically determines a set of actions that are executed on the network in order to achieve operational objectives (i.e., adaptation).

The work developed in 4WARD contributes to both parts of the control cycle. Specifically, we have developed algorithms for situation awareness that address real-time monitoring of network-wide metrics, group size estimation, topology discovery, data search and anomaly detection, of which examples are given in Sect. 8.3.

We have also developed algorithms for self-adaptation, briefly discussed in Sect. 8.4, giving special attention to self-adaptation in the context of other areas of research in 4WARD, such as VNets and GPs. Furthermore, we have developed self-adaptive solutions for the self-organization of the management plane, aiding configuration planning. Finally, also in the context of self-adaption, we have developed a framework for designing self-adapting network protocols inspired by models of chemical processes.

In order to support INM algorithms we have created an INM framework, which is discussed in Sect. 8.2. The framework is an enabler of management functions and it defines a set of architectural elements from which any distributed and embedded management structure can be created.

8.2 A Framework for INM

The 4WARD INM framework supports management operations in the future Internet by means of a highly distributed architecture. The main objective is the design of management functions that are located close to the management services, in most of the cases co-located on the same nodes; as target approach, they would be co-designed with the services. In this objective we identify the INM paradigm of embedding management capabilities in the network.

In line with a clean slate approach, the framework proposes the fundamental principles and constructs that state how to design and operate concrete networks according to the INM paradigm. This is the basis for defining algorithms in Sects. 8.3 and 8.4 as functions distributed in networks and, from them, to construct management operations through self-organizing mechanisms.

8.2.1 INM Principles

The INM framework stipulates five fundamental principles that will guide the design of management capabilities in the future Internet. These design principles are used as common ground for the design of the fundamental elements of the management framework. Additionally, the INM framework combines technical results with a methodology for a gradual, non-disruptive, adoption of the novel INM functions. The first principle addresses the basic ideas of the INM paradigm and captures all the potential developments of self-management features:

1. *Intrinsic principle*: Management is intrinsic to the network. This fundamental principle captures the fact that the network *is* the management entity at the same time. As such, this principle dictates all architectural considerations of the INM framework.

The following three principles are consequences of the intrinsic principle and support the design of embedded management functions in the future Internet. These principles are extreme cases that are generally relaxed in practical situations.

2. *Inherent principle*: Management is an inherent part of network elements, protocols, and services. As such, management functions come as inseparable, i.e. *co-designed*, parts of the network. For instance, in a structured peer-to-peer network, overlay management is implemented inherently by the peer-to-peer machinery and can be considered an inherent management capability.

3. *Autonomous principle*: This principle leads to the adoption of a fully self-organizing management plane, which would also automate the enforcement of high-level business goals and physical intervention. This principle is clearly not feasible in its pure form, but it defines long term objectives for research in automation of the future Internet: self-management functions will go beyond mere adaptation of device parameters and will strive toward the inclusion of domains traditionally excluded, such as network management guided by business objectives.

4. *Abstraction principle*: External management operations occur on the highest possible level of abstraction. In the theoretical extreme case, the network may be triggered by an external stimulus only once at the beginning of its lifetime. All subsequent management actions and processes are concealed and autonomous in the sense of the autonomous principle. This principle guides us toward the definition of management interfaces for operators that hide internal self-management processes more than today's approaches.

Furthermore, the INM framework defines the following architectural principle that addresses the gradual architectural design methodology:

5. *Transition principle*: The architectural design principles 2–4 should be implemented and developed in operative networks in a way that they can be gradually adopted. This principle is essential in that it allows gradual deployment of self-management 4WARD technologies and, in particular, assures marketability of INM results.

8.2.2 INM Transitional Degrees

While the architectural principles 2–4 are theoretic in nature, the transition principle breaks them down into a corresponding functional design space (shown in Fig. 8.1). This principle supports a gradual adoption of these principles to various practical degrees. In the center of the diagram, INM designates the extreme case where principles 2–4 are adopted in their pure form.

Along the *degree of embedding*, the INM framework provides scope for a relaxation of the inherent principle. Management processes can be implemented as external, separated, integrated, or inherent management capabilities of the network. Integrated is weaker than inherent in that instead of being indistinguishable management functionality, it designates visible and modular management capabilities, which are still closely related to and integrated with specific services. Separated

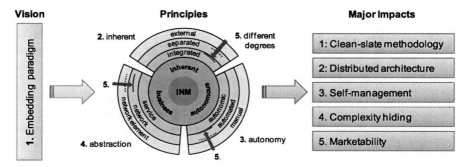

Fig. 8.1 Transitional diagram: three dimensions of the functional design space

management processes are those that are more decoupled from the service, for example weakly distributed management approaches. External management processes include the traditional management paradigms widely used today.

Along the *degree of autonomy*, the INM architecture allows for different degrees of autonomous management, from manual to fully autonomous processes. Manual refers to the direct manual manipulation of management parameters, such as manual routing configurations. Automated management can typically be found in the application of management scripts. Autonomic and autonomous degrees include intelligence that allows the system to govern its own behavior in terms of network management.

Along the *degree of abstraction*, different levels of management according to the telecommunications management network (TMN) functional hierarchy can be adopted. This dimension leads to a reduction in the amount of external management interactions, which is key to the minimization of manual interaction and the sustaining of manageability of large networked systems. Specifically, this dimension can be understood as moving from a *managed object* paradigm to one of *management by objective*.

As suggested by Fig. 8.1, different parts of the network may adopt their specific degree of embedding, autonomy, and abstraction based on practicability, specific goals, and other requirements. The INM principles guide the transition in the functional dimensions. If design issues are considered at the design time of new components, those components may encapsulate existing management functionality in a way that allows for a non-disruptive transition to purer cases of INM.

8.2.3 Architecture of the INM Framework

The INM principles and transitional degrees translate into the following four key design approaches that the INM framework follows when implementing network management:

Fig. 8.2 Main actors in INM

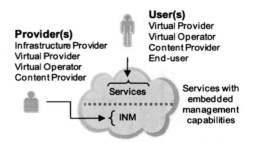

1. *Co-location* (structural): The INM framework enables the realization of management functions that are *co-located* with the services subject to management. This objective emphasizes tight integration from a *structural* perspective.
2. *Co-design* (functional): Complementary to co-location, the INM framework adopts a style of designing management functions in conjunction with service functions termed *co-design*, which emphasizes tight integration from a *functional* perspective.
3. *Collaboration* (functional, collaboration): In order to achieve highly distributed and self-adaptive operation of management functions, the INM framework provides clear definitions of how *collaboration* between building blocks of management functions, called *management capabilities*, is to be performed.
4. *Management by objective* (functional, organization): The INM framework follows the paradigm of *management by objective* rather than that of *managed objects*, defining how to enforce and report high-level objectives in a hierarchical manner by means of *management domains*, *self-managing entities*, and *management capabilities*.

Before introducing the main concepts for the modular and distributed architecture of INM that follow these guidelines, it is necessary to understand the main actors and how they relate. For simplicity, we refer to a model comprising two roles, which will help to identify the main beneficiaries of INM, like illustrated in Fig. 8.2:

1. *Provider*: Entity who is responsible for and operates network resources, either physical or virtual. The provider maps to different business roles like infrastructure and service provider as well as virtual network operator. Each will require INM to operate its respective resource domains. To the provider, INM offers a set of interfaces to operate those resources and to respect service level agreements (SLA's) with users.
2. *User*: Entity who requests and uses providers' resources as services. An entity that assumes the role of a provider can at the same time assume a user role when making use of another provider's resources as services.

Given the legal relationships between providers and users, an SLA is used to define the type of service delivered to the user and the guaranteed quality in delivering it. An SLA's technical description is therefore used as input by INM to configure services. Such descriptions can take the form of XML documents with well-defined negotiated network performances.

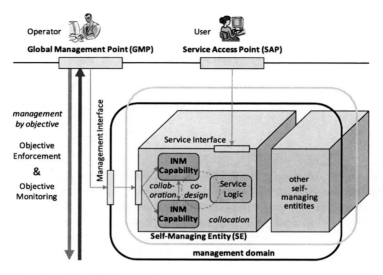

Fig. 8.3 Overview of the INM framework

Figure 8.3 illustrates the INM framework's main concepts. On the operator side, the *global management point* (GMP) is the high-level entry point into the management of a network. The GMP is the only management interface visible to the operator and provides the highest level of abstraction by means of objectives.

In the first hierarchical refinement, the global management point provides access to one or more *management domains*, each allowing access to a well-defined subset of the embedded management plane. Multiple domains may exist at any time during network operation, their configuration may change and they may be set up and torn down dynamically over time. Subdividing the embedded management plane of a virtual network may occur in both structural and functional terms, such as a subdivision by self-managing entities (see below) and the isolation of only performance-related management functions, respectively.

While the purpose of management domains is to extract a well-defined management subject in terms of structural and functional characteristics, *self-managing entities* (SEs) encapsulate self-management functions of individual services, such as networking of information. They are the logical constructs that encompass the properties necessary to achieve autonomous operation of a future Internet service-centric network infrastructure. SEs provide the means for embedding a set of generic properties that enable network operation with only high-level intervention from the operator. For that, SEs collaborate with one another and enforce objectives on the service level in order to meet the service-specific objectives dictated by the operator's high-level objectives. However, how this breakdown into lower-level objectives is performed is still an open issue.

Each self-managing entity contains multiple management capabilities (MCs), which implement the actual self-adaptive management algorithms on a fine granularity. All algorithms described in Sects. 8.3 and 8.4 are implemented at this level.

While containment of MCs within SEs is one possible realization, management capabilities may also exist stand-alone. This configuration is useful for management capabilities that encapsulate more general management algorithms, for example, generic monitoring functions. These possibilities reflect the degrees of embedding described in Sect. 8.2.2, and appropriate MC interfaces also allow the mix of different degrees of embedding in the same management system.

Each management domain combines a number of self-managing entities (SEs), more precisely, a structural or functional subset of a set of SEs. In other words, a management domain slices a set of SEs in such a way that limited management functionality is isolated, for instance to allow for an operator to only access a well-defined and security-enabled subset of management functions. The two most basic slices would correspond to the organization and collaboration interface of SEs, and more specific, slices may correspond to any of the self-management properties defined by an SE. Slices are indicated in Fig. 8.3 by multiple overlay planes intersecting the same set of self-managing entities. Note that slicing only applies to those degrees of embedding that are externally accessible via interfaces, that is, all except the inherent degree of embedding.

The enforcement and monitoring of objectives occur along the hierarchy of the previously described elements. Regarding the enforcement, each objective is specified on an abstract level by the operator at the level of the global management point and split into sub-objectives via management domains and self-managing entities until reaching individual management capabilities. On the other hand, objectives are monitored and aggregated toward the global management point in the opposite sequence of elements.

Figure 8.3 also illustrates how the INM framework addresses each of the four principal objectives. In particular, co-location and co-design operate between management and service functions, the latter being accessible via service access points (SAP's).

8.3 INM Real-Time Situation Awareness

Real-time monitoring of network performance and detection of anomalies and faults are critical for autonomous network management. While providing input for network operators, they also serve as a foundation for self-adaptation to changing conditions within the network, such as adaptive routing, search, and QoS management. We have developed solutions for, e.g. real time, adaptive monitoring [4, 6], decentralized threshold detection [16], and probabilistic network management [1]. Here, we provide four examples of distributed algorithms for monitoring a network in real-time: a comparison between gossip- and tree-based schemes to network measurement aggregation in terms of robustness and performance; an approach to distributed fault- and anomaly detection; a statistical scheme and algorithms which avoid network implosion resulted by a large number of nodes responding to a query or event; and a brief discussion on search challenges in self-organizing networks.

8.3.1 Algorithmic Aspects of Real-Time Monitoring

Providing real-time, continuous estimates of global metrics, such as the sum, average or maximum value of a parameter, over a large number of network entities is critical for efficient network monitoring and INM. An important aspect in the context of INM is the evaluation and comparison of different algorithmic approaches to distributed aggregation of such global metrics. This evaluation and comparison should be based on a number of key properties:

1. Performance under realistic, "normal" operating conditions, in the sense of accuracy and response time.
2. Controllability, in the sense of the ability to offset performance against overhead in a predictable manner.
3. Scalability. This property concerns the ability of the algorithm to handle network configurations of increasing size, without introducing unmanageable bottlenecks. Normally, this is interpreted as a requirement for sub-linear growth in overhead as a function of network size.
4. Robustness. The algorithm should be able to maintain functionality even under adverse operating conditions, including random faults (e.g. node failures), local overload conditions occurring in the network, and security attacks of various kinds.

We have focused on algorithmic aspects of the distributed aggregation of local measurements across the network. Specifically, we have focused on two main classes of algorithms, tree-based and gossip-based algorithms. We have addressed the following objectives: comparison of the two classes, their adaptation to various aggregation tasks, and evaluation with respect to the above comparison properties. Our findings are primarily based on:

1. Designs and evaluations of tree-based and gossip-based protocols that are resilient to random node failures, either fully or partially [16].
2. Designs and evaluations of tree-based and gossip-based protocols for distributed threshold detection [14, 16].

Overall, our theoretical and experimental investigations indicate that in a number of respects the tree-based protocols are superior to the gossip-based aggregation protocols we have been looking at, in terms of:

1. Overhead/accuracy ratios.
2. Ability to recover from random failures.
3. Ability to converge to accurate values in stable configurations. This is important for instance when cryptographic information needs to be aggregated, as is needed in, e.g. privacy applications.
4. Ability to support a wide range of aggregation functions.
5. Fast convergence.
6. Ease of analysis.

Fig. 8.4 Comparison of gossip-based vs. tree-based algorithms, in terms of robustness and performance. Figure shows estimation error vs. protocol overhead

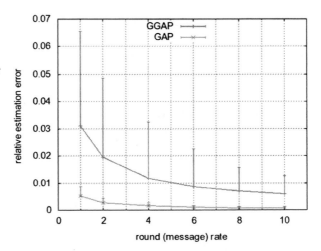

For instance, we consistently found close to a factor of 10 reduction of the parameters we estimated while passing between a gossip-based to a tree-based solution [16]. In Fig. 8.4, we plot the estimation error in terms of message rate for the robust gossiping protocol GGAP (Gossip-Generic Aggregation Protocol) vs. the tree-based aggregation GAP (Generic Aggregation Protocol) developed in earlier work, using a trace based on real network data [16]. Other experiments, e.g. measuring estimation error in terms of network size, failure rate, or protocol overhead, provide results that are consistent with this observation. This also applies when comparing our gossip-based and tree-based solutions for distributed threshold detection.

For the other points in the list 1–6 above we note first that there are standard, easy solutions available to make tree-based protocols robust against random node failures, at the cost of significant transient errors [3]. We found that the problem of making gossip protocols robust is far more challenging. Indeed, we are not aware of a robust version of gossiping with a comparable scope as, e.g. the basic GAP protocol. In fact, the solution we have developed is robust only to random failures that are non-contiguous in the sense that no two neighboring nodes of some given node is allowed to fail within a single round [15]. For the remaining points in the list 1–6 we can note that in general these are well-known standard points in favor of tree-based algorithms.

However, there are also significant arguments against tree-based approaches. This includes the propensity to concentrate management traffic to a few network links, and the sometimes dramatic transient errors which may arise as a result of tree reconfiguration in response to node failures. Also, we have encountered some indications that, at very high levels of churn, the performance of tree-based protocols begins to deteriorate so significantly that they begin to be outperformed by gossip-based ones. This point, however, is not yet understood well enough to make firm conclusions.

For the above reasons our conclusion is that tree-based approaches are preferable to gossip-based ones for network management applications, under reasonable assumptions on node failure rates.

8.3.2 Distributed Anomaly Detection

Within the 4WARD project for INM, a distributed approach to adaptive anomaly detection and collaborative fault-localization has been developed. The distributed algorithm is designed to statistically detect communication failures and deviations from normally observed response delays and packet drops on each link. The response delays and packet drop rates are obtained through distributed monitoring, where each node monitors adjacent nodes by sending probes. Compared to, e.g. the threshold-crossing detection capabilities of the GAP family of algorithms mentioned earlier, which can be used for coarse-grained anomaly detection purposes, the algorithm tries to localize detected anomalies to specific parts of the network. To pinpoint the origin of detected anomalous behavior to a node or link, the neighbors of adjacent nodes are used for collaborative fault-localization [12].

The distributed anomaly detection approach is focused on meeting the following three goals: first, the autonomous adaptation of algorithm parameters should significantly reduce the need for manual configuration. Second, the adaptability of the probing mechanisms should allow for improved efficiency of bandwidth usage, compared to, e.g. conventional monitoring based on fixed interval probing. Third, the algorithm should, without rigorous modification, run on different types of networks, network layers and data input.

The idea is to statistically model the expected probe response delay and packet drop rate for each connection of each node, and use the estimated models for parameter adaptation such that the algorithms network resource usage is adapted to local conditions. Instead of specifying algorithm parameters in time intervals for probing and specific thresholds for when traffic deviations should be considered as anomalous, manual parameter settings are here specified either as a cost or as a fraction of, e.g. the expected link latency or the probability of obtaining a probe response. Compared to monitoring methods with fixed probing intervals, the proposed anomaly detection algorithm continuously adapts probing intervals based on the locally observed probe response delays. This way the total link load caused by probing traffic can be reduced.

The approach to distributed anomaly detection is capable of adapting to long-term changes in observed probe response delays. The adaptation of observed probe response delays is achieved by learning temporally overlapping statistical models in a circular scheme, where the previous model is used as prior input to the next model. This type of learning mechanism includes temporally palimpsest properties and allows for smooth adaption to long-term changes, while gradually forgetting earlier observations.

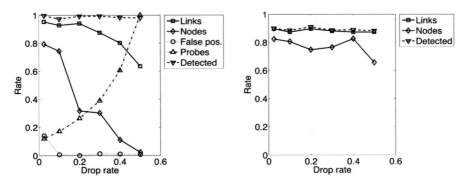

Fig. 8.5 Performance of detection and localization of communication failures (*left*) and latency variations (*right*)

The detection mechanisms are designed to increase the certainty of whether an observed behavior really is anomalous or not. For example, the detection of communication failures involves sending a series of probes for the purpose of increasing the certainty of the detection. The number of probes in the series and the interval between the probes are autonomously adapted to the observed probe response delay and packet drops (Fig. 8.5). In the case of detecting changes in the expected probe response delay, comparisons between current and previous response delay models are frequently performed, using the symmetric Kullback–Leibler divergence for Gamma distributions as metric [11].

Initial evaluation of the extended anomaly detection algorithm indicate that link and node anomalies caused by either shifts in expected latency or communication failures can be detected and localized to a certain link or node with satisfactory performance. In Fig. 8.5 the detection and localization performance is shown for increasing packet drop rate in the network. The experiments were performed in the OMNET++ simulator using scale-free networks as input. We see that more than 90% of the anomalous events are detected, and that more than 80% of the events of each type can be correctly localized to a node or link.

Due to the distributed nature of the approach, it is able to run on different types of networks and network levels (e.g. link level or service level) while scaling well to the network size, as the number of packets (including both control messages and probe traffic) scales linearly with the number of connections.

8.3.3 Adaptive Avoidance of Network Implosion

The "Not All aT Once!" (NATO!) algorithm [2] addresses a problem that arises in many modern networks, such as sensor networks, grid networks, satellite networks, and broadband access wireless networks. Such networks consist of thousands of end devices (nodes) that are controlled and managed by a single gateway. At times, due to a state change or a local event, a large group of end nodes must send feedback messages to the gateway.

The previously mentioned GAP algorithm can be used to, e.g. count the number of nodes in a group. NATO! is an alternative, statistical scheme for precisely estimating the size of a group of nodes affected by the same event without explicit notification from each node, thereby avoiding feedback implosion. The main idea is that after the event takes place, every affected node waits a random amount of time taken from a predefined distribution, before sending a report message (RPRT). When the gateway receives sufficient RPRTs to estimate the number of affected nodes with good precision, it broadcasts a STOP message, notifying the nodes that have not reported yet, not to send their RPRTs. This effectively adapts the number of RPRT messages needed to current network conditions. The gateway then analyzes the transmission time of the received RPRTs, defines a likelihood function, and uses a Newton–Raphson method to find the number of affected nodes for which the likelihood function is maximized.

Using mathematical analysis, we can provide upper and lower bounds for the estimation error at approximately $1/(N - 1)$, where N is the number of sent RPRTs, which is always over-estimated. This property can be used to bring the estimation error very close to zero. Simulation results show that with only 20 feedback messages coming from a group of 100 or 10,000 affected nodes, the estimation error is about 5 percent, and after error correction, the error is eliminated.

NATO! is applicable for networks and systems that meet the following requirements:

1. The network consists of a large group of end nodes reporting to a single gateway. Having each affected node sent a separate RPRT message to the gateway would result in one of the following implosion effects: (a) insufficient network resources for forwarding the messages to the gateway; (b) insufficient gateway CPU resources for processing all these messages; (c) delayed gateway response to the event.
2. RPRTs are identical (e.g. Ack, Nack); If this is not the case, the gateway should form a Broadcast START message with a query that is expecting such RPRTs. NATO! Is not useful when RPRTs contain data that is unique to each sender.
3. In order to correctly respond to the event, the gateway needs a good estimate of the number of nodes that have experienced this event.
4. The estimation should be completed in a timely manner.
5. The gateway is able to broadcast a STOP message to all of the nodes that might be affected by the event, to stop further transmission of their RPRTs.
6. The setup allows for precise timing, i.e., (a) the event occurs at the same time, or the server can start NATO! by means of a START broadcast message; (b) all nodes are time-synchronized, or the network delays are known.

As long as these six requirements are met, there are several INM applications for NATO!. As an example, INM QoS management can take advantage of the NATO! scheme. A cross-layer QoS management capability needs to collect information from all nodes within its domain, including available transfer rate, one-way delay,

BER, etc. However, this is not always possible, as not all nodes are integrating sophisticated mechanisms that are required to respond to such request. The NATO! scheme can implement an alternative approach; rather than collecting row information for QoS assignment, the QoS MC broadcasts a specific query, e.g. which nodes can accommodate a flow with specific QoS requirements. Only nodes that are capable of responding to such query, and can accommodate such request, start NATO! (for a delayed positive response). The NATO! gateway algorithm precisely estimates the number of nodes that can accommodate such request, without explicit notification from each one. With this information, it is possible to determine if QoS-aware routing is feasible for a given request.

8.3.4 Search in Dynamic and Self-organizing Networks

Search and routing in dynamic, self-organizing networks usually cannot rely on stable topology from which search tables, shortest paths and other optimized access techniques are derived. When no reliable indices or routing tables are available, flooding, random walks or gossip-based methods have to be considered to explore the network. These approaches can exploit partial knowledge available on the network nodes to reach a destination, but the search effort naturally increases with the lack of precise information due to network dynamics. This problem is especially relevant for wireless technology with strict limitations on power consumption. We address the efficiency of random walks and flooding for exploring networks based on case studies evaluated by simulation and transient analysis. In this way, performance tradeoffs are demonstrated when combining shortest path routing with randomized techniques. There are at least three networking scenarios, which lead to increasing dynamics when integrated into future Internet structures:

1. Peer-to-peer (P2P) and other self-organizing overlays.
2. Mobile ad hoc networks (MANET).
3. Sensor networks.

A search may refer to users, network nodes, information, content or services of any kind residing on network resources based on unique identifiers like IP addresses or hash values as often used in P2P networks. Developments on top of P2P systems have advanced toward distributed databases systems, which can be built in a scalable and efficient way even on unstructured network topologies and in the presence of unreliable nodes.

The *bubblestorm* approach [13] sets up a randomized replication scheme, where a data item is distributed on a set of nodes forming a data bubble. The size of each data bubble is kept proportional to the square root of the network size, which can be effectively estimated by the NATO! algorithm discussed above. The replication scheme achieves a high reliability and improves the performance of search as well as the throughput by enabling multi source downloads. Experience of the bubblestorm architecture in large networks has shown that a simultaneous disappearance of up to

90% of the nodes, e.g. caused by breakdowns in the underlying transport network, still leaves the remaining network and database system intact.

The search can be done by flooding a request or by random walks in order to exploit the self-organizing network. In flooding, messaging overhead is reduced restricting the flooding by a predefined hop limit h, which is set with regard to knowledge of the network structure and demands for coverage. Alternatively, a small initial value for h may be stepwise increased if the search radius turns out to be insufficient. In this case, however, a new step revisits all the nodes of the previous step, and as another disadvantage the number of nodes being reached for a larger search radius is not known a priory and is often increasing exponentially in unstructured or scale-free networks.

In this respect, randomized techniques are useful. A basic random walk exploits the network as a stepwise process, which proceeds from a node to a neighbor with the next edge in the topology chosen randomly. The main drawback of a single random walk is the long delay, as it may take a winding route through the network. This motivates us to propose new routing schemes for sensor networks or to combine random walks with flooding.

Combined variants include:

1. Random walks with an additional flooding step with small radius from all nodes being traversed or from the last node.
2. Starting several random walks in parallel or branching a random walk into multiple paths.

As a rigorous, self-adapting overhead control scheme, the number of messages in a random walk, flooding or combined search can be limited by a time to live counter, also denoted as *budget-controlled search* [5]. When the search is split up in multiple paths being traversed in parallel, the budget must also be split. Multiple random walks in parallel may be applied in a compromise between demands for low delay and low overhead.

A randomized replication of data as in the bubblestorm database architecture provides a favorable environment for random walk searches. Besides studies showing favorable properties of random walks in large unstructured networks, other investigations promote random schemes for ad hoc and sensor networks. In addition, it is known that random walks can efficiently exploit imprecise and only partially valid information in support of a search or when many nodes in the network are able to respond, i.e. when it is sufficient to reach one node in a larger set.

8.4 Self-adaptation Within In-Network Management

Using an INM approach to network management, the network itself must take proactive or reactive network management actions for the purpose of recovering a fault, avoiding a predicted fault, optimizing the network operation, or enforcing new or modified objectives submitted by the network operator. This is typically achieved by changing the network configuration, setup, or resource allocation. Based on the real-time monitoring and anomaly detection techniques discussed in the previous section, which are the basis for learning the current state of the network and its operation, we now turn our attention to how to make use of this knowledge in order to take corrective actions. In this section we discuss two specific examples of INM self-adaptation: how to ensure network stability under configuration changes and congestion control based on emergent behavior.

8.4.1 Ensuring INM Stability

As INM processes are running completely automatic, there is a need for configuration mechanisms during both the setup period and runtime. A fully automatic system might be unsuitable in certain situations, e.g. secure systems. However, manual interventions within the INM can be difficult to tackle as an automatic system can become inoperable by wrong configuration changes from an outside entity.

For these reasons a configuration planning module is required. A configuration component facilitates manual intervention with the INM, while a prediction component prevents unsound changes, thereby ensuring network stability. This configuration/planning module is considered to be running and available after the system has bootstrapped—during start-up the prediction module is unable to perform due to lack of real field data, and only new configurations can be applied via the configuration component.

Configuration planning is done by an outside entity called administrator. His interactions with the INM take place through an API, which allows him to check the system state and modify settings. Reconfiguration requests are processed by the configuration module, which invokes the prediction module for simulating the results of the proposed changes.

The prediction module implements a Markov chain of states, where a future state of the system depends only on the current state. The current network state comprises of all active nodes data containing:

1. Number of live (active) connections.
2. Load of each node.
3. Number of loaded connections (active users that use most of the bandwidth).
4. Average lifetime of a connection and delay of a package.
5. Alarms and errors (if there are any).
6. Current policies (e.g. high-priority route policies such as VoIP which are to be kept alive under any circumstances).

The prediction module runs as follows. Assume a set of given states in which the system can be. The system can go from one state to another with a certain transition probability. A stable network state is characterized by an equal or bigger number of live connections, at the same or a better connection speed while maintaining the current status quo between all network nodes. States are ranked in chronological order and computing the transition probability defines the probability of the system going from one stable state to another. The transition probability is the sum of all probabilities the system has any of its variables going in a different state. A transition matrix is constructed for each system variable. Nodes are considered to be independent of each other, and for each node a number of matrices are constructed.

A general system matrix is constructed, at the end, for each system variable. This matrix uses the probabilities computed for each independent node, and for each system variable. Thresholds are set up for every system variable that is of interest. If the thresholds get over or under run, that is if a smaller probability is detected, the prediction module reports its findings and the network administrator is notified.

This method allows for fast, on the fly and on demand simulations. It saves precious memory as no network history is needed. Because of the fast computing time, this mechanism is suitable for merging networks. It is also beneficial for mission critical applications as it allows for quick access to resources in order to stay alive while having a minimal impact on the existing hardware infrastructure.

8.4.2 Emergent Behavior-Based Congestion Control

The basic principle of congestion control used in today's Internet relies on the implicit or explicit collaboration of end systems and routers. It can be described as follows: in case the filling level of a router's queue reaches a pre-configured threshold, queue management strategies are applied to drop packets in a random fashion. TCP connections of end-systems affected by these artificial packet losses reduce their sending rate to avoid the impending congestion. In case of UDP based communication it is up to the affected application to implement a suitable flow or congestion control principle. In fact the required interaction between end-systems and routers can lead to an unfair sharing of network resources. To address such issues, we have focused on an INM congestion control approach, which does not rely on the interaction between end-systems and routers. With the aim to reduce required configuration and management overhead to a minimum, we applied a principle known as emergent behavior to congestion control, more specifically the emergent phase synchronization phenomena of pulse-coupled oscillators, which is a well-known phenomenon in biology and physics. To apply this synchronization property to congestion control we started by:

1. Associating each queue in a router with an oscillator based on the Mirollo and Strogatz model [9].
2. Identifying the filling level of a queue with the corresponding oscillators frequency (by using a linear function) [7].

Within 4WARD, the applicability of the approach has been studied in the context of Multipath-Routing Scenarios. In Multipath-Routing, multiple alternative paths between a data source and sink are calculated which can be utilized for the actual data transfer. In our approach, each path calculated, defined a group of oscillators associated with the corresponding router queues (e.g. g1, g2 and g3 in part (a) of

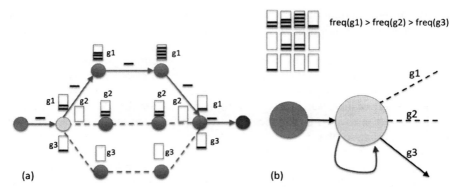

Fig. 8.6 Least congested path first scenario

Fig. 8.6). The oscillators in each group are assumed to be coupled during or after path calculation.

The idea behind the approach is that based on the synchronization property of the corresponding oscillators, the highest oscillator frequency in g1, g2 and g3 defines the frequency of the whole group. Consequently, in case of congestion as shown in part (b) of Fig. 8.6, the edge router is able to determine the path with the lowest maximal filling level for the next packet by comparing the frequencies of its oscillators.

As the result of our evaluation work we can state:

1. After an appropriate selection of oscillator related parameters the sync property emerges in the target multipath scenarios. We evaluated scenarios with a path length up to 100 routers.
2. For the addressed scenarios, the sync property also emerges in case of frequency variation. The resulting group frequency is the frequency of the fastest oscillator.

Thus, the resulting group frequency can be used as a basis for congestion control.

8.5 Relation to Other 4WARD Technologies

The work on INM presented here contributes to and have implications on several other areas of work performed within 4WARD. At a general level, 4WARD is exploring the development of a design process for combining existing or specifying new networks with customized architectures. INM fits into this design process within the *Governance* and *Knowledge strata*. The Governance stratum is related to a management domain and the Self-managing Entities that exist within it, and specifies the domains management objectives. The information produced by Management Capabilities is fed into the Knowledge stratum, which processes the information, presents it to operators, and feeds it back to the Governance stratum. Further, the management algorithms developed for INM serve as design patterns for solving different monitoring and adaptation problems. These algorithms can be key components of a design patterns repository and be included at the network design stage.

Network management functionality produces a significant amount of information, through counters, events/alarms and configuration parameters that determine system state during operation. 4WARD proposes a generic information-centric approach to networking, Networking of Information, where the identity of an information object is used as a reference to retrieve it from any node that might store the full information object. We have developed a preliminary set of guidelines on how to apply INM principles to networking of information [10]. As INM processes are implemented in a distributed manner and need efficient access and storage of information, networking of information is a suitable candidate for management information dissemination.

In the context of virtual networking, INM concepts apply in two places. First, INM provides management operations to the base infrastructure. Second, INM capabilities and functions are implemented in accordance with the network architecture running within a virtual network. Management capabilities in both places might need to interact, e.g. for the purpose of optimization or locating a network disturbance. INM potentially addresses several issues within a virtual networking setting. The decentralized operation of the infrastructure achieves a well-balanced resource usage and concludes possible virtual node re-location. Within the project, we have developed distributed reallocation schemes for virtual network resources [8]. Further, the virtual links are subject to resource management actions in order to guarantee the bandwidth and QoS requirements. Finally, the issues involved in root-cause analysis and fault management are of major importance in a carrier-grade virtual networking environment.

For Generic Paths, INM provides two key mechanisms: distributed real-time monitoring and resource adaptation. These mechanisms can be used for multiple purposes in path management, such as network state evaluation, distributed resource management, decision making for routing strategies, anomaly or fault detection, and fast failure recovery. INM also contributes to generic path routing strategies by providing information on the quality of routing paths, e.g. through providing accurate link availability statistics and through collaborative measurements, which differ significantly from simple path estimation approaches incorporated in present routing protocols. The generic path mechanism would collaborate with INM entities to provide support for real-time traffic path and resource availability monitoring, also across composite generic paths.

8.6 Concluding Remarks

As networks grow larger and more heterogeneous, INM techniques are likely to become significantly more important. This is driven by a need to lower manual intervention to reduce operational expenditure, in combination with a need for scalable solutions to manage ever larger and more complex networks. INM solutions, that are scalable and autonomous, will in the future find applications in all types of networked systems.

The main achievements and results of the work within 4WARD on INM are INM framework to support management operations and a set of distributed management algorithms. The INM framework serves as an enabler of management functions. By defining three main elements, namely management capabilities, self-managing entities, and management domains, complex management functions can be constructed and modelled within the framework. The project's prototype implementations illustrate key features that allow to enforce and monitor objectives and to induce self-adaptive behavior in collaborating management capabilities.

For estimating the network state, which is a necessary input to self-adaptation mechanisms, we have focused on a subset of the management tasks involved in situation awareness. Specifically, we have developed real-time monitoring of network-wide metrics, group size estimation, data search, and anomaly detection. Multiple aspects of self-adaptation have also been developed, including ensuring network stability under control changes and emergent behavior based congestion control. We have seen that robust, distributed algorithms can be devised for a multitude of management tasks without introducing excessive amounts of overhead in the networked devices.

References

1. M. Brunner, D. Dudowski, C. Mingardi, G. Nunzi, Probabilistic decentralized network management, in *Proc. 11th IFIP/IEEE International Symposium on Integrated Network Management (IM 2009)*
2. R. Cohen, A. Landau, Not All aT Once!, A generic scheme for estimating the number of affected nodes while avoiding network implosion, in *Infocom 2009 Mini-conference*, Rio de Janeiro, Brazil
3. M. Dam, R. Stadler, A generic protocol for network state aggregation, in *Proc. Radiovetenskap och Kommunikation*, Linköping, Sweden (2005)
4. A. Gonzalez Prieto, R. Stadler, Controlling performance trade-offs in adaptive network monitoring, in *Proc. 11th IFIP/IEEE International Symposium on Integrated Network Management (IM 2009)*
5. G. Haßlinger, T. Kunz, Challenges for routing and search in dynamic and self-organizing networks, in *Proc. 8th Internat. Conf. on Ad Hoc and Wireless Networks (Ad Hoc Now)*, Murcia, Spain. LNCS, vol. 5793 (Springer, Berlin, 2009), pp. 42–54
6. D. Jurca, R. Stadler, Computing histograms of local variables for real-time monitoring using aggregation trees, in *Proc. 11th IFIP/IEEE International Symposium on Integrated Network Management (IM 2009)*
7. M. Kleis, Sync: Towards congestion control based on emergent behavior, in *9th Joint ITG and Euro-NF Workshop Visions of Future Generation Networks (Euroview 2009)*
8. C. Marquezan, L. Nobre, G. Granville, G. Nunzi, D. Duduwski, M. Brunner, Distributed re-allocation scheme for virtual network resources, in *Proc. IEEE ICC09*, Dresden, Germany, 2009
9. R.E. Mirollo, S.H. Strogatz, Synchronization of pulse-coupled biological oscillators, SIAM J. Appl. Math. **50**(6), 1645–1662 (1990)
10. K. Pentikousis, C. Meirosu, M. Miron, M. Brunner, Self-management for a network of information, in *Proc. IEEE ICC09*, Dresden, Germany, 2009

11. R. Steinert, D. Gillblad, Distributed detection of latency shifts in networks, in *SICS*, Kista, Sweden, Rep. T2009:12 (2009)
12. R. Steinert, D. Gillblad, Towards distributed and adaptive detection and localization of network faults, in *Proc. of the 6th Adv. Intl. Conf. on Telecommunications (AICT 2010)*
13. W. Terpstra, J. Kangasharju, C. Leng, A. Buchmann, BubbleStorm: Resilient, probabilistic and exhaustive P2P search, in *Proc. ACM SIGCOMM 2007*, Kyoto, Japan, pp. 49–60
14. F. Wuhib et al., Decentralized detection of global threshold crossings using aggregation trees, Comput. Netw. **52**, 1745–1761 (2008)
15. F. Wuhib et al., Robust monitoring of network-wide aggregates through gossiping, IEEE Trans. Netw. Serv. Manag. (TNSM) **6**(2) (2009)
16. F. Wuhib, M. Dam, R. Stadler, A gossiping protocol for detecting global threshold crossings, IEEE Trans. Netw. Serv. Manag. (TNSM) **7**(1) (2010), doi:10.1109/T-WC.2008.I9P0329

Chapter 9
How Connectivity Is Established and Managed

Hagen Woesner and Thorsten Biermann

Abstract An architecture for data transmission that puts technological and administrative domains (compartments) in the role of the keeper of this shared information is described. Paths are established between communicating entities, basic functional blocks that reappear in different layers of the Internet. We explain how certain functions like routing, access control, and resource management are recurring in entities at all layers, and therefore allow an object oriented definition of entities and paths. Compartments and generic paths limit the scope within which state information needs to be kept consistent. Compartment layering is fundamentally different from the established ISO/OSI model and the chapter discusses several examples for the use of cooperation between more than the traditional two end points of a transmission.

9.1 Introduction

The main observation that triggered our work to develop a data transport architecture was the difficulty to introduce new functionality/protocols into today's network stacks. One reason for this is that there is no coherent way to identify communicating entities within the network and to manipulate them.

The network of today has grown to be a highly complex interconnection of transport technologies in which the only visible entities—Internet Protocol (IP) routers—are just one slice of which a particular user gets a partial view, at best. Underlying

H. Woesner (✉)
Technical University of Berlin & EICT, Berlin, Germany
e-mail: hagen.woesner@tu-berlin.de

T. Biermann
University of Paderborn, Paderborn, Germany
e-mail: biermann@ieee.org

L.M. Correia et al. (eds.), *Architecture and Design for the Future Internet*,
Signals and Communication Technology,
DOI 10.1007/978-90-481-9346-2_9, © Springer Science+Business Media B.V. 2011

connections, be it MPLS, Carrier Ethernet or optically-switched wavelengths only appear to limit the neighboring relationships between routers, reducing the complexity of this neighborship management (i.e. IP routing). This reduction of complexity, however, limits the degrees of freedom in the choice of paths in the network. Application gateways, firewalls, NAT and masquerading limit this choice even further.

A more flexible, powerful, and reusable data transport architecture is needed, and we base this on an approach to identify and manipulate data flows. This approach, however, needs to be generic enough to stretch over a wide range of technology levels and should encompass a wide range of data processing and forwarding functions in end systems as well as in intermediate nodes. Along with the flows themselves, we also have to cater for entities that manage, control, and realize these flows; examples are protocol engines in end systems or forwarding engines in a router.

With such a set of requirements, it is impossible to come up with *the, single* solution for a *one-size-fits-all* flow type. Therefore, we decided to choose a design process that combines (1) a uniform appearance and interface for all different flow types and (2) flexibility in supporting a wide range of flow types in as many different environments as possible.

We chose an object-oriented approach to design network components while keeping them coherent in their interfaces and basic structures. This allows to incorporate new networking techniques more flexibly than in today's network architectures as networks can be arbitrarily composed of the components. Furthermore, networks can easily be adapted according to any cross-layer information during runtime, thanks to the unified interfaces. Examples for data transport aspects that have been modified/integrated into our architecture are routing, mobility [1], cooperation and coding techniques [3], and resource allocation.

9.1.1 Flows and Paths

Communication at large scale typically does not take place back-to-back, but instead requires a number of intermediate nodes that switch circuits or forward packets along a—typically pre-established—path. *Routing*, the act of path establishment in a distributed system, is the result of neighbor discovery and the subsequent exchange of resource information. A path can be established with or without explicit signalling; any treatment of data other than best-effort usually requires signalling of a path identifier and the requested Quality of Service (QoS) parameters. It is beneficial to decouple *routing* from *forwarding* conceptually as well as technically, as the latter operation is a much less complex one. While the complexity of routing depends on the size of the graph and the frequency of changes of the available resources (due to, e.g., mobility), forwarding is determined by the nature of the path. A pure circuit switching does not require any evaluation of addresses whereas a connection-less datagram service like IP has rather complex longest prefix matching operations per packet.

A *flow* is a sequence of datagrams that share certain properties (like source/destination or QoS parameters). The size of a flow in the Internet follows a heavy-tailed

Fig. 9.1 Overview and interaction of GP architecture components. Entities are drawn as *rectangles*, CTs as *rectangles with rounded corners*, EPs as *diamonds*, GPs as *horizontal lines*, and Hooks as *vertical lines*

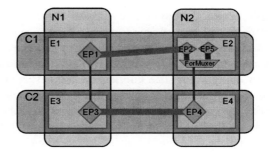

distribution but the general observation is that of mice and elephants. This means that there is a large number of short (even single packet) flows and a rather small number of very long flows. The size of the elephant flows has increased over the last years due to higher bandwidth in the broadband access links and bandwidth-hungry applications.[1] Evidence for this can be found in the weekly statistics of Internet2 [8]. This effectively means that a large number of packets follow the same path and that an optimization of the forwarding process for these packets would off-load a router from much of the per-packet processing it has to do in a connection-less datagram transfer.

9.2 Components of a Path-Centric Network Architecture

This section introduces the components that constitute the data transport framework: Compartments (CTs), Generic Paths (GPs), Entities, End Points (EPs), and Hooks. Two key concepts (CTs and Entities) have already been introduced earlier in the chapter on naming and addressing (which this chapter builds upon); nevertheless we repeat them here for the sake of completeness. An example of their interactions is given in Fig. 9.1. The scenario consists of five Entities (E1–E5), four CTs (C1, C2, N1, N2), four EPs (EP1–EP4) forming two GPs, and four Hooks. The GPs between EP1 and EP2 in C1 (e.g., IP) is realized by the GP between EP3 and EP4 in C2 (e.g., Ethernet). Entity E2 has two active Endpoints (E2 and E5) and implements the optional ForMuxer.

9.2.1 Entity

An Entity is the generalization of an application managing a resource. Depending on the implementation, this can be a process, a set of processes, a thread. The resource controlled by the Entity can be manyfold: think of an information object (web page,

---

[1] What used to be email is now Facebook, what used to be an MP3 file is now an HD video stream, all this with the size of an individual frame limited to the 1500 byte of the 1970s Ethernet.

video stream), the bitrate of an optical transmitter, even electrical energy or food can serve as picture of what is meant by resource here.

An Entity is controlling and managing the resource, whereas the actual access to it—communication—is provided by the Endpoints of a GP.

An Entity keeps state information that is shared among multiple GPs and runs processes or threads that manage this state. Examples for such state information are routing tables, resolution tables, and access control tables. In a traditional networking model Entities belong to the control plane, while the actual generic paths go into the data plane. The fundamental difference here is that the impetus for communication does not come from an 'application above' but indeed from the Entities themselves. This however still means that one Entity may need the service (a resource) offered by another Entity. This can be exemplified by bandwidth that is required for the transmission of a video stream. While one Entity controls the video stream (it can be source, sink, or both) it needs underlaying Entities to transport the stream. So one resource may require another.

9.2.2 Compartment

A Compartment (CT) is a set of Entities that fulfill the following requirements:

- Each Entity carries a name from a CT-specific name space (e.g., MAC addresses in the Ethernet CT). These names can be empty and need not be unique. Rules how names are assigned to Entities are specific to each CT.
- All Entities in a CT *can* communicate, i.e., they support a minimum set of communication primitives/protocols for information exchange. These protocols are implemented as different GP types. Hence, for joining a CT, an Entity must be able to instantiate the EP types required by the CT.
- All entities in a CT *may* communicate, i.e., there are no physical boundaries or control rules that prohibit their communication.

A special CT is the Node CT (N1 and N2 in Fig. 9.1). It corresponds to a processing system, i.e., typically an operating system that permits communication between different processes (e.g., by using Unix domain sockets). By means of virtualization, multiple Node CTs can be created on one physical node.

An Entity is typically member of at least two CTs, the "vertical" Node CT and a "horizontal" CT. Furthermore, the Entity has a (possibly empty) set of names from each of the respective CT name spaces.

9.2.3 Generic Path

A Generic Path (GP) is an abstraction of data transfer between communicating Entities located in the same or in remote nodes. The actual data transfer, including forwarding and manipulation of data, is executed by EPs.

9.2.4 Hook

A *Hook* is a GP within the node compartment. It implements the binding between two Endpoints or between an Endpoint and a ForMuxer, if that option is used. Besides exchanging data, Hooks also hide names from other CTs to permit changing a GP's realization later on. A Hook number is equivalent to a port, but the Hook itself is more than the port number, in that it has two identifiers (at its two ends). These identifiers (= Hook numbers) are local to the respective Entity. The reason for keeping two identifiers instead of one is the easier change of bindings. For an Entity, essentially, the local Hook number remains the same when the binding of this Hook is changed to another Endpoint.

9.2.5 Endpoint

An End Point (EP) keeps the local state information of a specific GP instance, i.e, it is a thread or process executing a data transfer protocol machine and doing any kind of traffic transformation. EPs are created by an Entity and may access shared information of that Entity.

Usually, GPs require other GPs to provide their service. E.g., a TCP/IP GP requires another GP that provides *unreliable* unicast, like an Ethernet GP, to provide a *reliable* unicast service at the end. Therefore, EPs are bound via Hooks to other EPs within the same node.

Note that GPs cannot cross CT boundaries due to the possibly different name spaces, protocols, etc., GPs always reside within a single CT.

9.2.6 Optional: ForMuxer

Forwarding and multiplexing is an inherently simple function consisting of three steps. At first, an *input arbiter* decides among the available ports where to fetch the next data. Second, based on a combination of some internal state and names (labels) transported in a frame header a Forwarding Information Base (FIB) is consulted that provides a rule about the outgoing port. And third, the data frame is put into the according output queue. In an architecture that splits between control and data, *formuxing* would clearly go to the data path, which means that it can be delegated into EP functionality. This would mean that there are multiple Hooks going in and out of an Endpoint, which then performs forwarding decisions in addition to writing or removing header information. The second option is to come up with a dedicated building block, the *ForMuxer*, that explicits the forwarding and multiplexing. This block corresponds to a dumb switch controlled by the Entity and attached to the Endpoints via the data path, i.e., Hooks. Having a separation between the Endpoint and the ForMuxer allows for cleaner separation of concerns, and hence, interfaces. This clean separation can mean that all functions that modify the header of a frame are executed in an Endpoint, while the ForMuxer only forwards. When considering the case for a Time to Live (TTL) decrement (and the following recomputation of

Fig. 9.2 Mediation points
are a union of several Entities.
The data path, connecting
Endpoints using a single
ForMuxer, is not shown

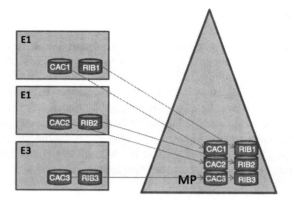

the header checksum) within an IP router it becomes clear that the thorough split of *processing* and *forwarding* is not always reasonable.

9.2.7 Optional: Mediation Points

A Mediation Point *maps to* an Entity and a corresponding data processing and forwarding part. It is an Entity that belongs to many CTs and has the capability to "mediate", i.e., switch, between the respective Endpoints. Technically (mapping-wise) it can do so by converting all attached data formats into a single and controlling one large *ForMuxer*. To the outside (the Config level) this appears as one point where arbitrary GPs can be merged, interstitched, and converted. To do so, Mediation Points (MPs) need to be member in one CT that spans across Entities in client nodes and creates a "unified control plane", an approach that can be found in GMPLS today. The need for this Über-CT is evident when considering the situation of two Entities who are member in different CTs and that is: name spaces, for example, in a LAN using 48-bit addresses as names and an telephone network. The Entity in the LAN will not be able to communicate with its counterpart in the other CT as long as one does not join the CT of the other or both join a common, say, Session Initiation Protocol (SIP) CT. An MP therefore is *one option* of implementation, as it may be seen in Fig. 9.2.

MPs have pros and cons, and have led to huge discussions along the track of the work within 4WARD. One can briefly say that the trade-off is between cross-CT optimization on one side and the flexible introduction of new CTs and scalability on the other. An MP can be modelled as a union of several Entities (that delegate their control functionality into one) with a number of attached Endpoints.

9.3 Mapping of Functions into the Architectural Building Blocks

The following sections present one way of using the GP architecture to actually establish communication between Entities. We distinguish and map certain functions into blocks that are responsible for controlling, processing, and formuxing data. A separation of control and data into "planes" can be cleanly executed this way, allowing to pinpoint the "what goes where".

9.3.1 Endpoints and Entities—State Information Keepers

The establishment of any communication produces state information within the communicating parties. This state information belonging to an individual communication relation (the GP) is contained in its Endpoint. There is, however, state information that is to be kept outside of the Endpoint, as it relates to the establishment of new and the management of existing GPs. This information is attributed *control*, and we give it a placeholder: the Entity. So both Entities and Endpoints are state-keepers, one in the control, and one in the data "plane". Entities not only create and manage Endpoints, but insert and retrieve data through them, as well. In fact, the API of the GP is executed between an Entity and its Endpoint(s). There may be more than one Endpoint controlled by an Entity, corresponding to GPs to different destinations within the CT and/or different services. There is a similarity between a GP and a Forwarding Equivalence Class (FEC) used in Multiprotocol Label Switching (MPLS). Entities and Endpoints live within CTs and have no control or understanding of names or data formats outside of their own CT. Entities are named, and, by previous definition, the CT defines the space in which names make sense. Endpoints have identifiers that are referenced locally to the Entity.

9.3.2 Endpoints and Entities—Data Processing and Control

In addition to the state information there are the processes creating and managing this information. This means that processes related to transferring data and the control of the data flow are placed in the Endpoints. As an example, there will typically be one state machine per, say, TCP connection that controls errors and the flow between two Entities. So the separation in a TCP/IP CT is between an Entity (having an IP name) and one or more TCP or UDP Endpoints controlled by this Entity. Endpoints are identified by their port numbers, the functions specific to the Endpoints being:

- Error control
- Flow control (TCP congestion control being a combination of both error and flow)
- Header processing
- Data manipulation (encryption, (trans-)coding, network coding, etc.).

Processes in Entities regard the control and management of GPs, and that is, their respective Endpoints. Functions to be executed in Entities are

- Service Discovery
- Routing
- Name resolution
- Access control
- Management of records of the resources used by the GPs.

9.4 Prerequisite Mechanisms to Set up Generic Paths

Necessary prerequisite for the creation of a generic path is the existence of an Entity (being source or sink of information). When the Entity is created (for example, a *driver* for a network card or a Web server is installed/started), it creates an initial GP that plugs into the backplane of the node CT, which is assumed to be a logical bus.[2]

The layering introduced in the GP architecture is reflected by repeated execution of three steps in order to create a GP.

1. Service Discovery. The service of another CTs has to be matched to the service required by the GP that is to be created. If one or more CTs offer a matching service, then the Entity within these CTs can be used to create GPs in a next level. The service will in its simplest form define a common data exchange format.
2. Routing table lookup. The Entity checks the internal routing table (or the one of the MP) for a next hop *within its CT*. If there is no next hop, the destination is assumed to be resolvable locally.
3. Name Resolution. The next hop is resolved into a set of addresses, i.e., the bindings it has to other CTs.

This recursion stops whenever an Entity finds that it already has a GP established to the desired destination, in which case it returns the handle to it. In many cases this will be the PHY layer, which will return when a link is established. It may as well be any higher Entity that appears to reach the desired destination with a required service (for example the connectivity discovery would usually return at IP for any higher application).

9.4.1 GP Service Discovery

An Entity will be able to create a certain set of GPs, i.e., it will create the corresponding EPs. Every of these EPs is described by the tuple <required_service, offered_service, offering_Entity>. This *service advertisement* is published through the backplane logical bus to all Entities of the node CT, where again centralizations and optimizations are possible. A GP Service Graph (GP graph) may then be constructed that consists of vertices and edges, where the service descriptions

---

[2]The MP option makes a different assumption of this backplane as being a switch.

are vertices and the potentially usable GP/EPs are the edges of the graph. The GP graph, bearing similarity to other routing graphs, allows shortest path routing, which reduces the number of processing steps or *layers* that data would have to pass. The graph—again it may be centralized in an MP or distributed in the Entities of the node CT—serves to find all Entities to which a specific Entity could bind in order to create connectivity. Note that the GP service graph is not necessarily saying anything about reachibility of a destination within a certain CT.

9.4.1.1 Resource Description Frameworks

The service graph shown in Fig. 9.3b is just one way of performing *service composition*. A simpler way of doing the same is the OSI layer model, where the resulting service graph would be taking the shape of the well-known hour-glass. Using an MP, the resulting service graph would be a star. OSGi frameworks as used for service composition of netlets within 4WARD are an option, though typically the service offered by OSGI *bundles* is different from GPs. An ontology of GP services using approaches like eXtensible Markup Language (XML) and Web Ontology Language (OWL) allows a generation of links in the graph even on imprecise or incomplete specification. This call is a *reflection interface* in that it is answered with a further set of parameters that describe the resource. The bind() call is handled by the Access Control process of the Entity, which gives the opportunity to limit the visible set of parameters in accordance to some credentials provided by the calling Entity.

At the end of the service discovery an Entity *binds* to another. It does so by creating a acGP to the initial (the control) Endpoint of the Entity that exported the required resources.

9.4.2 Neighbor Discovery and Routing

The GP service graph may contain several edges (options) leaving one vertex. This means that several Entities (in potentially different CTs) offer a specific service. Nothing is said so far about the actual reachability of the destination within these CTs. The binding between Entities now means that the two are *reachable through each other*. Therefore this binding is to be inserted in a CT-wide name resolution system. In the next step the Entity needs to find out if the newly established binding actually serves the purpose of reaching other Entities of the same CT. To give an example, it makes little sense for an IP router to add a Bluetooth interface if that is not connected anywhere, i.e., no other IP Entity can be reached through it. The similarity of neighbor discovery and routing has been explained in Chap. 5 on naming and addressing, here one can add that an Address Resolution Protocol (ARP) call may actually reveal existing bindings in the "lower" CT to the CT that the calling Entity belongs to. This means that neighbor discovery may start by sending an almost empty WhoHas packet to a broadcast address in both the upper and lower CT. This packet would be forwarded by the next Endpoint to other nodes in the CT

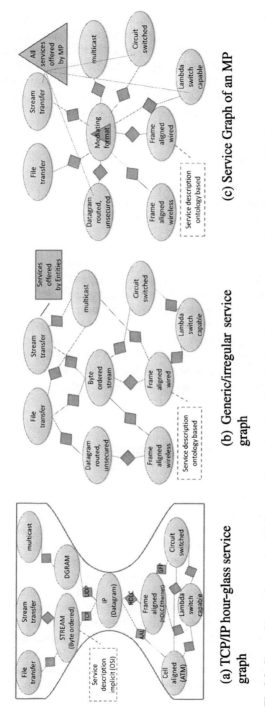

(a) TCP/IP hour-glass service graph

(b) Generic/irregular service graph

(c) Service Graph of an MP

Fig. 9.3 Shape of the service graphs of different architectural options. Note that the *rightmost* graph has a minimum diameter of 2 while the *leftmost* is maximum, resulting in a high number of *layers*

Table 9.1 Routing table of an entity E_1 in a compartment C_1

Destination entity	Via neighbor	Cost
E_2	–	15
E_3	E_2	23
F	H	15

and answered by all nodes in the lower CT who have a binding to any Entity of the higher CT. These answers reveal names and binding of other members of the same CT. Using these, the new Entity can publish or request a name. Typically, self-assigned names will not contain any structure that can be exploited for the purpose of routing, whereas topology information (indicating, e.g., a subnetwork or provider prefix) is assigned from the CT to a new Entity. The generic formulation "through the CT" leaves the individual way of name assignment intentionally open, but Dynamic Host Configuration Protocol (DHCP) may serve as an example. The Entity will, after it has a name, start exchanging information about its available resources, in other words, send out and receive advertisements for the state of these resources, construct a database of this resource information and execute a routing protocol on top of this database. This will produce cost metrics for all routes and populate the routing table (see Table 9.1).

9.4.3 Name Resolution

We refer the reader again to the routing and name resolution tables discussed in Chap. 5 on naming and addressing. However, and this extends the above discussion, we earlier introduced *Hooks*, explicit bindings between Endpoints in a Node CT. These Hooks can now be stored in a resolution table as well. In current name resolution schemes the port number is not explicitly stored along with the address, but the *well-known port numbers* tell implicitly where to leave the lower Endpoint from. Explicit storage of Hook numbers can thereby reduce the danger that comes from scans at known ports.

When resolving a name, an Entity needs to know the name of its desired peer Entity and the CT to which itself and this name belongs. The objective of name resolution is to find the following additional information for such a name:

- The name of a CT via which the peer Entity can be reached (e.g., WLAN SSID #4322).
- An Entity inside this *lower CT*, which can handle the communication on behalf of the originator Entity (typically, by means of sharing a Node CT).
- The name of a remote Entity in the lower CT that can pass on data to the actual peer Entity (typically, by means of sharing a Node CT).
- The "way up" from the remote Entity in the lower CT to the destination Entity in the upper CT. This is one of the identifiers of the Hook that is attached to the remote lower Endpoint.

Table 9.2 Name resolution table of an Entity E_1 in a compartment C_1

Entity	CT	down-Hook	Entity	CT	up-Hook
E_1	C_1	1234	E_3	C_2	5678
E_2	C_1	1345	E_4	C_2	5789

The core point in designing a unified name resolution system is to avoid spreading knowledge of how to interpret a name outside of its CT. Neither does the upper CT understand names of the lower CT nor vice versa. The only binding between the two is a Hook that has two identifiers that are again local to the Entities.

Hence, the only thing an Entity can do to resolve a name (in absence of further knowledge) is to contact all other Entities in its own CT and ask which Entity *has* this name (optimizations will come later)—a WhoHas message is sent *inside* its own CT. The subsequent IHave message is then evaluated to populate the name resolution table as in Table 9.2.

Examples discussing ARP and a P2P-based name resolution scheme, as it is used by NetInf, can be found in [4].

9.5 Establishing Connectivity

In order to actually create a GP an Entity needs to know the name of the remote Entity, the service required from that GP and an Entity in the node CT that would provide connectivity. This is a first assumption, and it will be shown later how the recursive process of service discovery, routing and name resolution leads to the knowledge of this triple. For now, let us look at an example of a Web client/server communication: In Fig. 9.4 a simple sketch of a client node $N1$ with two network interfaces, a wired (say, Ethernet), and a wireless (say, WLAN) is shown on the left. The Entity in the URL compartment ("browser") is requested to open a GP to a web server in order to retrieve a html file. The type of the Endpoint it will employ for this is chosen to be http (as opposed to gopher or https). With the request the Entity receives the name of the server (www.4ward-project.eu), usually this is actually imposed by some higher name resolution scheme (e.g., a search engine).

In a first step the browser Entity would need to check if it has an existing GP of the required type to the destination, in which case it would return a hook to this Endpoint to the calling application (which is an Entity in another CT). If not, the creation request is stored and a new GP is created in the following steps:

1. From the implementation of the http service to be offered by this Entity the `<required service>` for http is described, in this case ("ordered byte stream"), leading to a localization within the service graph. From there, the Entity can retrieve hooks to other Entities that provide this service. It finds that there is an Entity offering the service (whose name is 1.2.3.4 in the Host_CT).

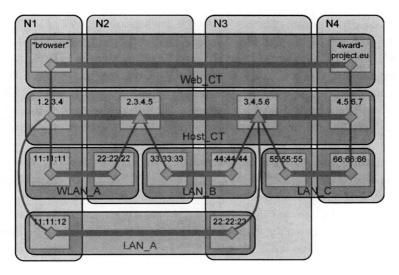

Fig. 9.4 Realization of an HTTP GP

2. The browser Entity consults its routing table of the URL compartment, but as the world wide web is one flat network there is no routing table currently in web applications.[3]
3. After finding that the "next hop" is the destination the web name needs to be resolved into an address in a lower CT. This is nowadays being done using DNS. We however assume a more generic resolver function as discussed above in Sect. 9.4.3. Assume for simplicity that the resolver mechanism and its location is known (through an entry in a node-local /etc/resolv.conf) and that the Entity knows how to contact this. It will return one or more CTs and names of Entities therein. (Note that these are indeed addresses for Entities of the Web CT). For all matches between the returned CTs of the service graph and the name resolution inquiry the creation of GP continues recursively. As name resolution returns *Host_CT*, 4.5.6.7, *hook*80 the browser Entity will next contact the local representant of the Host_CT, namely, 1.2.3.4. It will contact this Entity using its local name, known in the node CT, and request the creation of a GP to the destination 4.5.6.7 with the type "ordered byte stream".

Entity 1.2.3.4 offered this service as it can create TCP Endpoints. The procedure continues by finding the required service, say "frame aligned", and this time we find two Entities offering the same or similar service (see the ontology discussion in [14]), one 11 : 11 : 11 in the WLAN_A and one 11 : 11 : 12 in the LAN_A CT. The routing table lookup for the destination 4.5.6.7 results in two entries with different cost, say 2.3.4.5 with cost 5 and 3.4.5.6 with cost 4. The name resolution process for

---

[3]http redirects arguably have the same function. One might however think of a hierarchical (e.g., Top Level Domain-based) routing even in the Web.

both of the next hops will return that 2.3.4.5 is reachable through 22 : 22 : 22, *hook*1 in WLAN_A and 33 : 33 : 33, *hook*1 in LAN_B; 3.4.5.6 though 44 : 44 : 44, *hook*2 in LAN_B, 55 : 55 : 55, *hook*1 in LAN_C and 22 : 22 : 23, *hook*1 in LAN_A. There are two matching CTs here, LAN_A and WLAN_A, meaning that the creation of the GP can actually continue through one or both 11 : 11 : 11 and 11 : 11 : 12 Entities depending on the capability of the Endpoint. We assume that there is a multi-path TCP capable Endpoint available that can actually meet the correct forwarding decisions.

The creation of the GP now continues at the LAN compartment Entities with the request of a "frame aligned" GP to 22 : 22 : 22 and 22 : 22 : 23, respectively. Since both Entities find that there is already an established GP to the destinations they return hooks to it back to the calling Entity 1.2.3.4. The recursion within the node CT has stopped, and the creation of the GP continues within the Host_CT by passing on the previously stored createGP (dest. 4.5.6.7, type "ordered byte stream") command to the next hop(s). As the local hooks have been returned by the name resolver, the respective LAN Entities can address the Entities connected in the Host_CT directly. In fact, the Entities in the LAN compartments are not involved in any routing or forwarding decisions now, as the hook itself is enough information to address the attached Endpoint in the Host_CT. The Entities 2.3.4.5 and 3.4.5.6 can be considered special, as the respective Endpoints are "forward-only", which is why these Endpoints are depicted as MP.[4]

The same procedure with service discovery, routing table lookup and next hop resolution continues to the destination Entity in the Host_CT, where a TCP server listens behind the hook 80. There are some technical subtleties of creating multiple TCP Endpoints that look to the outside as one, but these are implementation details of today that are actually caused by the use of *well-known ports* and existing firewalls. In fact, the distinction of the TCP Endpoints today takes both the source and destination ports into account for demultiplexing. This is a matter of the specific IP Entity, but even this can be described using the proposed architecture.

There are two important things to note here: One is that there is no way of distinguishing between a GP creation request coming from "above" or "below". This reflects the fact the layering of ISO/OSI is dissolved into a generic service composition problem. The second is that the *semantic overloading* of IP addresses had to be resolved here into host and interface names in order to allow for multi-path selection.

---

[4]This distinction is helpful for a cleaner separation of forwarding and multiplexing from processing, see the discussion in Sect. 9.2.6.

9.5.1 Generic Path API

9.5.1.1 Creating a New GP

Creating a new GP in an Entity is as simple as calling `createGP()` with suitable parameters. This function needs at least four parameters: the CT in which the GP shall be created, the own Entity name, the name of one or more remote Entities, and the requested GP type. In addition, since it is usually required to get informed when the GP creation has been finished, a callback function can be defined. It will be used to inform about the status (success, failure) and the local side's EP. The methods for creating a GP are summarized in Fig. 9.5.

9.5.1.2 Using a GP

Now, after a new GP has been created, the Entity usually intends to interact with the new EP according to the communication paradigm the EP implements. In the following example, an Entity maintains two EPs of different types; one EP of a stream GP and one EP of a GP implementing the publish/subscribe communication paradigm. How the Entity uses these two EPs via their EP Application Programming Interfaces (APIs) is illustrated in Listing 9.1.

9.5.1.3 Modifying an Existing GP

The EP API does not only allow to communicate via a GP, it also permits to modify it. Specifically, the EP API allows to inspect properties of the GP, like its current data rate, resource information, or references to other GP elements (e.g., EPs, MPs) that are involved in realizing this GP. Furthermore, the API permits to configure the GP, like changing its error control behavior.

Using this EP configuration API allows to arbitrarily traverse the whole GPs structure that is involved in realizing a GP. This way, it is possible to inspect and reconfigure all elements on this way, like EPs or MPs, as long as their policies permit it.

9.6 Managing Connectivity

Changes in connectivity can arise from three main factors: congestion, fading channels in wireless networks and mobility of end users or sessions. In an abstract way, all these factors make an Endpoint observe oscillations in the capacity of the path. Reactions to this observation should be fundamentally different in every situation, though. In the following we show how the GP helps identifying the cause of the problem and how it may help mitigating it.

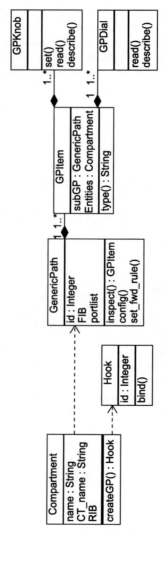

Fig. 9.5 UML class diagram of Compartment and Generic Path class. Note the recursive composition of GP

```
 1    public MyEntity : public Compartment {
 2        // ...
 3        Endpoint::pointer streamEP;
 4        Endpoint::pointer pubSubEP;
 5
 6        void someMethod () {
 7            // ...
 8            // Send data via stream GP
 9            streamEP->send (&data);
10            // Receive data from stream GP
11            streamEP->receive_blocking (&data);
12
13            // Publish data via publish/subscribe GP
14            pubSubEP->publish (&data);
15            // Subscribe to certain data via publish/subscribe GP
16            pubSubEP->subscribe_blocking (&data);
17            // ...
18        }
19        // ...
20    }
```

Listing 9.1 Using a GP for communication by interacting with its EP

9.6.1 Multipath Routing and Cooperative Transmission

The process of service discovery does not necessarily establish bindings between Entities that create a stack, but rather an irregular graph, leading to potentially multiple paths. The hierarchical naming and addressing that the GP architecture mandates makes multi-path forwarding an option at all CTs in contrast to approaches like HIP and LISP [5] which insert a single identifier/locator split. It also allows the insertion of special-purpose CTs for cooperation of a limited number of nodes in order to achieve certain goals. An example for this introduction of *domains of cooperation* is the Cooperation and Coding Framework (CCFW).

9.7 Cooperation and Coding Framework (CCFW)

Cooperation and coding techniques are usually applied to provide users with a better performance than plain forwarding of data can provide. Unfortunately, these techniques are not always beneficial, i.e., sometimes plain data transmission achieves the same or even better performance. Hence, cooperation and coding schemes have to be dynamically enabled in beneficial situations and disabled again when users suffer compared to plain transmission. To decide whether operating in cooperative or in plain mode is beneficial or even if a specific technique can be applied at all, a lot of information regarding the network topology, traffic passing a node, available resources, and interface properties are required.

Integrating cooperation and coding techniques into real systems is a complex task. To extend a single device with additional features to support just one type

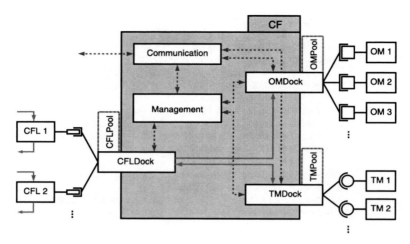

Fig. 9.6 Overview of the CCFW components in a classic layered networking system. *Solid arrows* denote data exchange; *dashed arrows* represent control connections

of cooperation or coding, modifications deep within the protocol stack are required which usually have to be done from scratch. This is true for the operations applied to the payload as well as for the already mentioned environment monitoring features. However, many of these functional units are required by many different cooperation and coding techniques and could be used in common to ease developing and prototyping new schemes and to avoid redundancy in the overall system.

We developed an architecture that aggregates functionality which is required by many different cooperation and coding techniques in a *Cooperation and Coding Framework (CCFW)* [3]. This framework is available at all nodes that participate in any cooperation or coding operation. Operation-specific functionality is encapsulated in separate modules. These modules report their specifications, i.e., environmental requirements and consequences of activating, to the framework. From then on, the modules are automatically activated in beneficial situations.

9.7.1 Components of the CCFW

The CCFW consists of four different components: the Cooperation/Coding Facility (CF) and various Observation Modules (OMs), Transformation Modules (TMs), and CF Layers (CFLs). A schematic of the data and control flow in between is shown in Fig. 9.6. More details can be found in [3, 4].

The central part of the CCFW is the CF. It is available *once* per node and controls all activities of OMs, TMs, and CFLs available at this node. The CF is split up into several functional units, each of them responsible for a certain task, like connecting the modules (the three docks), enabling/disabling certain modules (*Management*), or providing basic communication functions to modules (*Communication*).

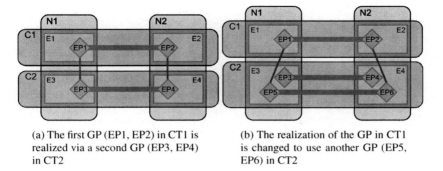

(a) The first GP (EP1, EP2) in CT1 is realized via a second GP (EP3, EP4) in CT2

(b) The realization of the GP in CT1 is changed to use another GP (EP5, EP6) in CT2

Fig. 9.7 Implementing traffic transformation with a special GP. The existing realization of the top GP is changed during its run time

OMs monitor a certain parameter of the node or of its environment, e.g., the link utilization or the neighborhood topology. Based on these observations, the management unit decides whether activating a certain cooperation/coding technique is possible and beneficial.

TMs provide the actual traffic transformation, i.e., the implementation of the cooperation/coding techniques. The TMs are instantiated on demand by the CF's management unit.

Accessing today's layered network stack requires to augment existing layers with additional cooperation/coding functions or to insert additional intermediate layers to redirect data to appropriate TMs. These layers are the CFLs, shown in Fig. 9.6.

The mapping of all these functions into the GP architecture and the resulting benefits are discussed in the following.

9.7.2 CCFW in the GP Architecture

9.7.2.1 CF Layers (CFLs) and Transformation Modules (TMs)

There are several possibilities to implement the CCFW's Transformation Modules (TMs) in the GP context. The obvious way to transform traffic into another format is using a special GP class for this. I.e., this specialized GP does additional coding/cooperation operations. There are two ways of using this mechanism. First, the new GP can be instantiated within an existing CT. Within this CT, the new GP simply replaces an existing one. Figure 9.7 shows this implementation.

In the second approach, the new GP does not replace an existing GP. It is instantiated in addition (in a new CT) and is then inserted in between two existing GPs that currently realize the end-to-end transmission (without the desired coding/cooperation feature). This technique is illustrated in Fig. 9.8.

Both of the described mechanisms have the advantage that switching between the operation modes is transparent for the upper end-to-end GP (EP1 ↔ EP2). On the

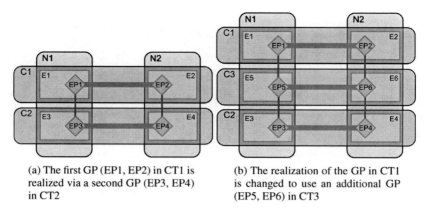

(a) The first GP (EP1, EP2) in CT1 is realized via a second GP (EP3, EP4) in CT2

(b) The realization of the GP in CT1 is changed to use an additional GP (EP5, EP6) in CT3

Fig. 9.8 Implementing traffic transformation with an additional GP. The existing realization of the top GP is changed during its run time

other hand, they cannot be applied arbitrarily, i.e., the scheme that inserts a new GP in an additional CT (Fig. 9.8) requires the implemented cooperation/coding technique to be independent of the underlying GP. E.g., wireless cooperation, which requires changes to the Medium Access Control (MAC) protocol, cannot be implemented like this. The advantage, however, is that independent cooperation/coding techniques, like encryption, can be applied universally in various situations. An example would be inserting the same encryption GP between the application and transport GP in one situation and between transport and link GP in another situation.

9.7.2.2 Management

The management unit of the CF controls all activities, i.e., collects input of various OMs and, based on this, activates the TMs whenever beneficial. As this requires information from many different GPs and CTs it is advantageous to not put these management functions into individual entities but into the ForMux control MP. The MP knows about all entities within its Node CT and maintains the hook table. This permits to decide which parameters have to be monitored and allows to transparently change the realization of GPs by adjusting the hook table.

9.7.2.3 Observation Modules (OMs)

To be able to decide whether a specific cooperation/coding technique is beneficial the management unit in the core needs information about the current networking environment. Such information, like packet error rates, link utilization, or neighborhood topology, must be gathered from several locations within the own node and from other nodes.

An obvious source for network-related information are the entities and EPs in the GP architecture. They are directly involved in communication and can provide information, like link utilization or channel quality. The information can be accessed via the Entity and EP APIs.

Another source for (remote) information is the In Network Management (INM) infrastructure, described in Chap. 8 of this book. It provides mechanisms to efficiently collect and distribute network parameters, which makes it a suitable base for the CCFW's observation activities. Besides standard information like network topology that is provided by INM, the GP-related information about entities and EPs can be fed into and distributed by INM as well. This simplifies the development of OMs as basically only one single OM is required—the one that uses the INM interface to gather information.

Using the CCFW architecture has major advantages compared to today's cooperation/coding-unaware systems. First, development and deployment of new cooperation and coding techniques is drastically accelerated in existing networks and in prototyping environments due to the CCFW's commonly available functions and its well-known interfaces. Furthermore, modules that are developed for a certain target environment, e.g., XOR coding in wireless context, are also beneficial for other scenarios, like wired networks. This results from the fact that the decision to activate a module is only based on its abstract specifications and an abstracted view on the environment, not on the actual physical environment.

9.8 Three Ways of Managing Mobility

In this section we present three distinct proposals that provide mobility concepts within the GP architecture framework [14]. These solutions take advantage of the object oriented properties of EPs for the dynamic update of GPs, thereby going beyond current concepts based on centralized mobility servers like in Mobile IP (MIP) based solutions.

These mobility mechanisms may coexist or may be combined dynamically according to the mobility context.

9.8.1 Dynamic Mobility Anchoring

The Dynamic Mobility Anchoring (DMA) approach provides a distributed, flat architecture able to support mobility in the heterogeneous network, see [14]. This is driven by the expectation that distributed control and data plane architectures achieve better performance and scalability than centralized architectures where central entities introduce much more delays, bottlenecks and constraints [2]. In DMA, mobility support is inherently sustained by Access Nodes (ANs) that act as traffic flow anchors, whereas the rest of the network remains unaware of mobility. When a multi-interface terminal moves in a heterogeneous network, it establishes a single

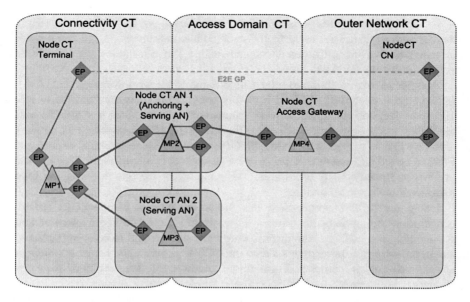

Fig. 9.9 The Dynamic Mobility Anchoring (DMA) scheme

or multiple network connectivity associations with one or more ANs. When set up, each flow is implicitly anchored on the initial serving AN, which then provides the necessary indirections through ANs to which the Terminal is moving. ANs actively cooperate to guarantee the delivery of the traffic flow within the access network by constantly considering the most appropriate traffic flows' mapping over available connectivity associations, adapting traffic flow indirections. Hence, the terminal can take benefit of any connectivity technology available depending on radio resources and Quality of Service fluctuations.

Figure 9.9 illustrates an example where a multi-interface Terminal uses a multi-path routing GP to transfer the traffic flow over its two active network attachments, its serving ANs (see AN1 and AN2 in the figure). AN1 acts also as the anchor for the considered end-to-end GP. It aggregates the traffic flow from/to the current serving ANs and the Access Network, allowing flow delivery over either AN1 or AN2 access interfaces. A MP is utilized to split or aggregate the traffic flow within the terminal or the anchoring AN. When the terminal moves out from AN1, its current GP's anchor remains located in AN1 until the GP is terminated. However, any new GP is then anchored on a serving AN, which takes benefit of DMA being dynamic to select the most appropriate anchor at each GP setup. Compared to existing mobility schemes, new dynamic and distributed properties introduced with DMA allow flexible and optimized mobility support. Traffic indirections is activated only when needed with local support in ANs, whereas more traditional scheme rely on systematic indirection between centralized (and eventually hierarchical) "static" mobility anchors (e.g. the Home Agent (HA) in MIP) in core networks and the terminal or its AN.

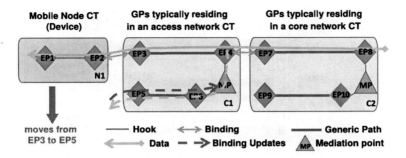

Fig. 9.10 The Anchorless Mobility (AM) scheme

9.8.2 Anchorless Mobility Design

The AM concept is based on the use of dynamic bindings between EPs, see Fig. 9.10. Having no specific mobility anchor (HA in MIP, [13], [7]), this concept solves the main mobility pain-point today, namely of scalability. Furthermore, the AM concept allows a local break-out, which is not easy to accomplish by using a tunneling solution like MIP. Despite there are possibilities depending on the mobility protocol used, the problem is that trombone routing can occur readily for signaling and data paths if no precautions are taken.

The proposed concept fits for all types of mobility: session, terminal and network mobility. Conceptually, in the GP architecture there is no relevant difference between inter and intra-technology handover, as modifications and reconfigurations of the GP are performed via a common API. Moreover, the AM concept uses different addresses for locations and for host identifiers in the network. By making use of a proper addressing scheme we also overcome the locator/identifier split [5] problem of today's Internet.

According to the object oriented nature of the GP architecture [14], different classes of EPs are derived from the EPs base class. A mobile device for example contains two different kinds of EPs, see EP1 and EP2 in Fig. 9.10. A logical EP at the Information Object (application) is wired to an EP denoting the device connectivity. This latter EP is connected via a Binding to an EP in the access network, denoting the current location of the device. Such an EP is visible from outside its surrounding CT and also internally connected to other EPs at the edge of this CT via GPs. Thereby end-to-end connectivity within the CT and thus throughout the network is ensured.

The mobile device is represented by a node CT. The access network represented by another CT constitutes the name space and as said before the internal connectivity of CTs is realized by GPs. The main component that facilitates mobility is the dynamic Binding function performed between the EPs. If for example the mobile device is moving, a Binding to another EP is established, before destroying the Binding to the former EP (softhandover). Note that the Binding always connects EPs of different CTs, while the connectivity inside a node CT is realized by Hooks, cf. Fig. 9.10. Upstream packets from the mobile device easily travel on the new path

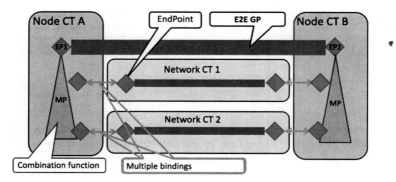

Fig. 9.11 Multihomed End-to-End GP

through the CT immediately after a re-binding to the new EP. Downstream packets from the core network toward the mobile device must be redirected at a suitable point in the network, e.g. controlled by a MP. Usually this point where the former and the new path reconvene lies on the former path and as close to the mobile device as possible.

In case of session mobility, the application is instantiated at a new device, and the state of the Entity (representing the running application) is transferred from the old device to the new device. The mobility of entire access networks, e.g. of a LAN in a train, is accomplished in a very similar manner to the mobility of a mobile device. This is because the involved node CTs and network CTs have the same functionality and APIs and are thus treated in a similar manner.

9.8.3 Multi-homed End-to-End Mobility

The design of a future multi-homing protocol for multi-interface mobile terminals requires the use of multiple paths between the communication terminals. These paths can be combined to provide composite services as one unique higher level end-to-end path [14]. Multihomed End-to-End Mobility (MEEM) is a mobility management mechanism defined for the GP architecture in which mobility is handled by the multihomed terminals.

As illustrated in Fig. 9.11, users of multi-interface mobile terminals can connect to different networks simultaneously through several interfaces. A Multihomed End-to-End GP (MEE-GP) is thus composed of several end-to-end sub-GPs. Mobility related to these interfaces can be handed in an end-to-end manner taking advantage of multihoming. When mobility occurs over an interface, seamless handover is much easier achieved by switching traffic into a secondary interface before the handover and back to the initial interface afterwards. This approach is especially suitable for the case in which the handoff decision is taken by the terminals.

This proposal introduces two functions for mobility management which are implemented within the MP, combination function and multiple binding.

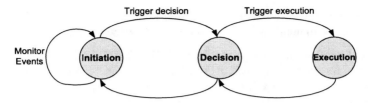

Fig. 9.12 Handover steps

- Combination function estimates performance of the combined end-to-end GP in order to decide the set of paths that should be used simultaneously and the scheduling algorithm that should be used over these paths to provide the best performance.
- Multiple binding sets up and maintains the bindings between the EP of the end-to-end GP and the EPs of the end-to-end sub-GPs. This binding is dynamic. When a node's locator is changed due to mobility, multiple binding is responsible to update the bindings with the new locator.

The advantage of end-to-end mobility management is that no new network entity needs to be defined. However, this approach should not be used in case of large number of end-to-end-path connected to a terminal (e.g. one of the terminals is a big server which has to maintain hundreds or thousands of connections from clients). End-to-end mobility should be used for applications such as video or audio streaming, video or audio telephony or other applications in which data transmission is carried out between two or several multi-technology multi-homed mobile terminals. End-to-end mobility management can be found in SCTP [15] and SHIM6 [11]. The MEEM principle can be applied to other existing protocols such as TCP or UDP to obtain multipath TCP or multipath UDP with end-to-end mobility support. It can be also used as a principle to design new protocols integrating multi-homing and end-to-end mobility management.

9.9 Triggers and Handover Decision

Three steps can be differentiated for supporting mobility management, and more precisely, handover management while maintaining ongoing communications for a moving End-Point (see Fig. 9.12):

1. **the handover initiation** results from information or event gathering (link monitoring, new incoming call, etc.), leading to a handover decision-making process
2. **the handover decision** has a direct impact on the network resource management, it selects a new connectivity for a dedicated communication taking into account user preferences, operator policies, access link QoS, etc.
3. **the handover execution** may include a possible handover preparation (pre-attachment, QoS context or service renegotiation, context transfer, etc.) before

the execution itself consisting of the End-Point connectivity switch and corresponding location update for the data forwarding adaptation.

Current wireless networks technologies (3G, WLAN, LTE, ...) apply different approaches for distributing such events, information and control functions among terminals and networking entities, at various protocol layers. In this context, the GP mobility framework needs to be highly flexible to self-adapt to various heterogeneous connectivity technologies on one hand and to provide efficient multi-level (terminal, session, network) mobility on the other hand. The realization of handover initiation, decision and execution functions in the three ways of managing mobility can be hence distributed and coordinated between hosts and one or several network entities:

- in DMA scheme, handover steps are coordinated in both terminal and access nodes entities, either being controlled at the terminal or the network side
- in the AM approach, different network entities can take part in the handover process, depending on the considered EP location (access network, core network, host)
- in the MEEM, the different handover steps are under host control.

Radio access technologies specificities, such as 3G or WLAN for example, become constrained to the wireless link CT only. Our three mobility schemes apply a global approach able to take benefit of the capabilities of various access technologies.

Beyond the pure handover execution performance, handover steps analysis outlines that an adequate handover triggering and decision framework is key for the global efficiency of mobility management schemes. Several frameworks to facilitate such triggering exist. In the Ambient Networks project a mobility toolbox as a component of the Ambient network architecture was proposed [12]. Also [16] enables to handle mobility mechanisms based on the modular extensible Ambient control space. Reference [9] defines a framework that supports the event collection and processing and triggering from hundreds of different sources. Lastly, the MIH model, see [6], support distribution of event and control information relative to handover initiation and decision steps. It allows for terminal's, network's or shared handover decision model as shown in [10].

All these frameworks concentrate on context information and triggering and do not refer to the mobility algorithms applied. We herein focus more on mobility execution functions, considering their use with such existing frameworks. Thus, in the following, we evaluate and compare more deeply our three mobility solutions.

9.10 Conclusion

This chapter gave a brief introduction into an architecture that due to its clean and recursive naming and address binding structure allows multipath routing, mobility of sessions, hosts, information objects, or users, in short: resources. The individual way of transporting informations or accessing the resources is left to the implementer, that is, connection-oriented or connectionless transport, or the scope of this transport (a wireless link or and end-to-end connection, all can be captured with a Generic Path approach).

References

1. P. Bertin, R.L. Aguiar, M. Folke, P. Schefczik, X. Zhang, Paths to mobility support in the future Internet, in *Proc. IST Mobile Comm. Summit* (2009)
2. P. Bertin, S. Bonjour, J.M. Bonnin, Distributed or centralized mobility? in *Global Telecommunications Conference, 2009, GLOBECOM 2009* (IEEE, New York, 2009), pp. 1–6, doi:10.1109/GLOCOM.2009.5426302
3. T. Biermann, Z.A. Polgar, H. Karl, Cooperation and coding framework, in *Proc. IEEE Future-Net* (2009)
4. T. Biermann et al., D-5.2.0: Description of Generic Path mechanism (2009), Project deliverable
5. D. Farinacci, V. Fuller, D. Meyer, D. Lewis, Locator/id separation protocol (lisp) draft-farinacci-lisp-12.txt (2009), http://tools.ietf.org/html/draft-farinacci-lisp-12. Expires: September 3, 2009
6. IEEE 802.21, IEEE standard for local and metropolitan area networks, part 21: Media independent handover services (11 November 2008)
7. D. Johnson, C. Perkins, J. Arkko, Mobility Support in IPv6, RFC 3775 (Proposed Standard) (2004), http://www.ietf.org/rfc/rfc3775.txt
8. A. Karp, S. Shalunov, B. Teitelbaum, Internet2 weekly netflow statistics, http://netflow.internet2.edu/weekly/
9. J. Makela, K. Pentikousis, Trigger management mechanisms, in *Proc. Second International Symposium on Wireless Pervasive Computing*, San Juan, Puerto Rico, USA, 2007, pp. 378–383
10. T. Melia, A. de la Oliva, A. Vidal, I. Soto, D. Corujo, R.L. Aguiar, Toward IP converged heterogeneous mobility: A network controlled approach, Comput. Netw. **51**(17), 4849–4866 (2007)
11. E. Nordmark, M. Bagnulo, Shim6: Level 3 Multihoming Shim Protocol for IPv6, RFC 5533 (Proposed Standard) (2009), http://www.ietf.org/rfc/rfc5533.txt
12. E. Perera, R. Boreli, S. Herborn, E. Jochen, M. Georgiades, E. Hepworth, A mobility toolbox architecture for all-IP networks: An ambient networks approach, IEEE Wirel. Commun. Mag. (2008)
13. C. Perkins, IP Mobility Support for IPv4, RFC 3344 (Proposed Standard) (2002), http://www.ietf.org/rfc/rfc3344.txt. Updated by RFC 4721
14. S. Randriamasy et al., D-5.2: Mechanisms for Generic Paths (2009), Project deliverable
15. R. Stewart, Stream Control Transmission Protocol, RFC 4960 (Proposed Standard) (2007), http://www.ietf.org/rfc/rfc4960.txt
16. V. Typpo, O. Gonsa, T. Rinta-aho et al., Triggering multi-dimensional mobility in ambient networks, in *Proc. 14th IST Mobile & Wireless Communications Summit*, Dresden, Germany, 2005

Chapter 10
How to Manage and Search/Retrieve Information Objects

Septimiu Nechifor

Abstract We present the overall vision for a network of information, illustrating the fundamental ideas, and explaining the mechanisms currently under development that will bring about a major paradigm change in networking. After briefly reviewing relevant scenarios where the current host-centric approach to information storage and retrieval is ill-suited for, we introduce how a new networking paradigm emerges, by adopting the information-centric network architecture approach. We illustrate how information retrieval may look like in the future, emphasizing on the user perspective. We then put forward the architectural requirements for a network of information. A description of the mechanisms, the "nuts and bolts" so to speak, of the technologies that implement a network of information is provided. We describe a network of information operation, providing examples and highlighting performance improvement. A long-term view is taken, with a discussion on evolution.

10.1 Introduction

The overall goal on the work with Networking of Information (NetInf) concept is to develop and evaluate communication architecture for an information-centric communication paradigm. NetInf should not only provide large-scale information dissemination, but also accommodate non-dissemination applications, including inter-personal communications; it shall inherently support mobile and multi-access devices, capitalizing on their own resources (for instance, storage). NetInf shall make it easy and efficient to access information objects without having to be concerned with or hampered by underlying transport technologies. NetInf shall also provide

S. Nechifor (✉)
Siemens, Brasov, Romania

L.M. Correia et al. (eds.), *Architecture and Design for the Future Internet*,
Signals and Communication Technology,
DOI 10.1007/978-90-481-9346-2_10, © Springer Science+Business Media B.V. 2011

links between the physical and the digital world. The architecture shall offer an interface to an application-neutral communication abstraction based on information objects. The information model shall make it possible to manage these objects in a secure fashion. The architecture shall be designed in such a way that a NetInf system is largely self-managing, see Fig. 10.1.

NetInf extends the concept of identifier/locator split for information management. Another level of indirection and the possibility for recursive lookups are added in order to decouple information from their storage location(s). In today's Internet information is assumed to be valid because the sender appears legitimate. In NetInf, hosts take a secondary role and as information ascend into center stage and pieces of data have to become self-certifiable. Users can focus on the actual information content instead of having to focus on their locations domains, as is done today, e.g., with URLs. NetInf addresses current problems such as unwanted traffic, denial of service, and intermittent connectivity by use of cryptographic identifiers, and by introducing an information-based communication abstraction which borrows ideas from the publish/subscribe paradigm. It is important to point out that NetInf is not an application-layer overlay and that it is a choice made in order to facilitate the integration of other new technologies in a new networking architecture. Examples of such new technologies developed within 4WARD include virtualization, generic paths and in-network management.

Internet was developed as a computer level communication abstraction, being implemented in a client–server manner, for HTTP by example. According to this paradigm, it is all about connecting network terminations, whereas the information exchanged on the channel is just a side concept. This approach has proven to be successful and useful for many purposes since its introduction. However, it is the way the Internet is used that has changed, consisting now most of the time in the retrieval of some kind of information.

The Network of Information (NetInf) research aims to address the issues uncovered by the current communication paradigm, from data security to flash crowds and DDOS attacks, from the complex operation of multicast and mobile communications that were not natively supported, to the lack of effective support for data dissemination, by developing a new communication architecture based on an information-centric paradigm.

In NetInf, users and client applications ask the network for a piece of information or a service, identified by its name and/or attributes, and the network satisfies the request. NetInf provides native support for content publication and retrieval, searching and storage services, and is intended as an enabling technology for new advanced concepts, like the server-less and semantic web. Data integrity can be embedded in the information itself and the communications can be made more reliable by exploiting opportunistic transport, content locality, caching strategies and network storage. Multicast and broadcast distribution of content among heterogeneous receivers can be efficiently implemented, thanks to network-based caching and storage capabilities. Moreover, as NetInf extends the concept of identifier/locator split, mobility of data and nodes can be easily supported.

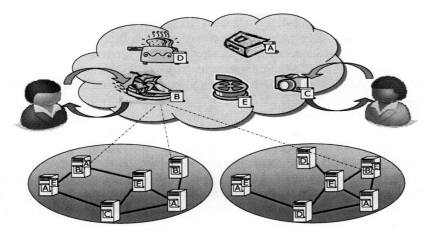

Fig. 10.1 Information centric networking

10.2 Information—A User Perspective

Information retrieval is dominating the usage of communication networks. Stored content in the form of music, video and computer software is typically transferred from servers with large storage capacities to PCs and mobile phones. Traffic stemming from live content, such as TV and radio, is continuously growing. Current networking technology is however increasingly being recognized as ill-adapted for this usage as well as being burdened with the ever increasing problem with security.

Current networking technology is based on a *device-centric* paradigm, which focuses on the interconnection of devices, such as computers, mobile devices, servers and routers. The information objects (as pieces of data) themselves are lacking identity independently of the devices they are stored on. The prevailing naming schemes, where DNS host names are part of the information object names, including URLs, effectively tie the information objects to the hosts (devices). The dominating method of transferring information using an end-to-end TCP connection makes the information more or less anonymous to the devices it passes through. The anonymity of the information objects makes it hard to cache them to avoid future unnecessary transfers. All of this, together with the lack of a global multicast mechanism, makes it hard to implement efficient information distribution. Up to now these limitations of the initial Internet design have been addressed mainly by overlays and interception of end-to-end flows. This leads to an increasing vulnerability and increased management cost beside the inherent inefficiencies of overlay solutions. Furthermore, new applications like IPTV and sensor networking demand more efficient as well as easy and consistently managed information network architectures as described in this document.

Information-centric networking is considered to be the third generation of communication networks. The first generation, the telephone networks, is about interconnecting wires, enabling users to have conversations with each other. Even though the technology has changed, the telephony paradigm is still that the end-devices are connected as if there were a physical wire between them. Second generation networking is about interconnecting devices, enabling services on those devices to communicate. The (new) third generation is about disseminating information, making that information available to applications and users efficiently on a large scale [1].

The need for information-centric networking is manifested by the increasing number of overlays that are created for the purpose of information dissemination (e.g., Akamai CDN, BitTorrent, Skype, and Joost). The objective of some of these solutions is to distribute information by relying on users to exchange pieces of data between themselves, massively distributing the load away from any central server, and scaling automatically to any group size. The NetInf architecture integrates much of the functionality of these overlays, including caching functions. NetInf extends the networking of information concept beyond "traditional" information objects (e.g., web pages, music/movie files, streaming media) to conversational services like telephony, and store-and-forward services like email. Special attention is paid to how this affects wireless communication and to how services can be made to work in an environment with a heterogeneous and disruptive communication infrastructure. Furthermore, we are extending networking of information to include real world objects, and by these enabling new types of services.

One of the challenges of the information-centric approach is to design a naming scheme for information objects and a name resolution system where the object names can be resolved in order to subsequently access the objects. The naming scheme needs to be designed with self-certifying identifiers such that authenticity and integrity of the objects can be provided without depending on the trust of the host delivering the object. Such a naming scheme can be seen as taking the notion of a identifier-locator split a step further to not only apply to hosts, but also to the information objects. Several models for identifier-locator separation have been proposed, e.g., the Host Identity Protocol (HIP) [2], the Internet Indirection Infrastructure (I3) [3], the Layered Naming Architecture [4] and the NodeID architecture [5]. By building on this prior work, we are able to design a networking architecture where mobility, multi-homing and security are an intrinsic part of the network architecture rather than add-on solutions. It will also allow users to gain increased control over incoming traffic enabling new possibilities for defending against denial of service attacks.

The information-centric communication abstraction has a number of advantages. Efficient distribution of content to a large set of recipients can be implemented. Content caching is built into the architecture and is thus provided without resorting to either interception of requests or special configuration at the receiver. Load-balancing is provided without depending on add-ons such as DNS round robin. Both reliability and performance are improved, since information can be retrieved from the closest available source.

Performance and reliability can be enhanced by an information-centric paradigm in a heterogeneous wireless environment where there are disruptions in communication, transient access opportunities and multiple access choices. An information-centric communication abstraction gives more flexibility in the delivery of data objects compared to using an end-to-end byte-stream. The network has better knowledge of the intent of the applications and therefore has the possibility to treat the data more intelligently. The network can easily deliver the data using multiple routes, redundancy over the available paths, and intermediate storage to overcome connectivity disruptions. The extreme is a scenario when an end-to-end path never exists [6]. With an end-to-end reliable byte-stream, the network has to violate the assumptions of the abstraction, or the application must implement these functions itself, including application gateways at suitable network locations.

The performance and reliability benefits from using storage also benefit non-dissemination applications, such as personal email. Another example is that direct delivery of email between two laptops with WiFi connectivity can be supported without involving infrastructure.

The information-centric approach gives new possibilities to prevent denial-of-service (DoS) attacks. With an information-centric approach nobody can force network traffic your way without your consent. This is made possible by moving control from the sender to the receiver. A sender can make information available, but for that information to be transferred, the receiver has to ask for it. Prevention of DoS attacks is a motivation to design NefInf as a replacement to the current TCP/IP for end-to-end communication.

10.3 Architectural Requirements

Figure 10.2 illustrates the NetInf architecture and its components from the view of a NetInf node. Application programs use the functions provided by the basic *NetInf Application Programming Interface* (NetInf API), and/or the more advanced functionality provided by a *NetInf additional service*, for example, storage (NetInf Storage API). Using the basic interface, the application can publish and retrieve information and data objects, both static objects and channel objects which provide a stream of information. The basic API can provide not only a copy of an object to the application program, it can also provide a handle to a suitable transport entity via which the object can be accessed (read or write access)—in the simplest example, such a handle could even just be a socket with a TCP connection to a server where the object is hosted.

There are some distinct conceptual engines in the architecture. The *name resolution engine* in the middle resolves object identifiers into locators where copies of the corresponding object can be found. That process can involve consulting the node-local resolution engine and/or using one or several name resolution protocols to query resolution engines external to the node. The local resolution engine keeps a directory of objects present on the local node, both ones that are temporarily cached by the cache engine, and ones that are managed by the local store engine. Resolution requests can come either from the API or from the network via one of the resolution protocols. The local storage engine manages the objects published through the basic

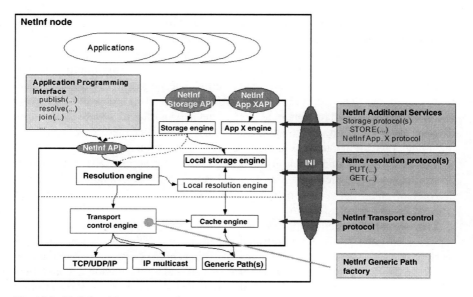

Fig. 10.2 NetInf architecture overview

NetInf API and objects are stored locally as part of a storage service. The local storage engine can be seen as an interface to file systems and other means of persistently storing data.

Resolutions requests normally map object ID to locator, but can also use object meta-data to perform look-up. In case the lookup keys are ambiguous and don't match meta-data used in the resolution engine the resolution fails. The name resolution engine can then instead invoke a search engine to try to use the semantic properties to reason about meta-data relations and map them into object ID. This object ID can then be resolved into a location. Search engines can be provided as NetInf additional services.

The *NetInf additional services* level is on top of name resolution. It provides optional add-on services which may depend on additional business agreements. The additional services can be operated by third-party service providers, and there might be multiple providers for the same service. An example is the storage service providing persistent storage. Information is stored using the NetInf Storage API, but can then be accessed using the basic NetInf API. The additional services works together with what is identified as the basic NetInf API, for example when an IO is published, an associated storage action is invoked.

The *transport control level* is below name resolution. The transport control engine coordinates which protocol to use for the access of NetInf objects and the protocols used internally in NetInf. The NetInf transport control protocol is needed to coordinate transfers with other nodes in the network. The protocol uses an addressing and routing scheme (explained in the Evolution section with the MDHT mechanisms) which is independent from the protocol that performs the actual transfer. The cache engine manages objects temporarily stored on the local system. It is similar to the local storage engine, but any object in the cache may be removed at any time to make space for other objects. Objects in the cache can be delivered to local as well as remote consumers. The transport control engine interacts with the cache engine to be able to transfer objects hop-by-hop in order to overcome disruptions in connectivity. The transport control level can engage functions of the Generic Path abstraction; for example, it can act as a Generic Path factory to generate instances of suitable GP classes that are tailored to interact with the transport control and caching engines.

The transport control level can be seen as an overlay when existing standard technology is used for the transfer, for instance, HTTP, or other protocols that run on top of TCP/IP end-to-end. This possibility is important for migration to NetInf technology. However, to fully utilize the advantages with the NetInf paradigm, the lower protocol stack can implement or call tools for specific purposes, like to provide measures against denial of service attacks. Such an approach could profit from the design framework and concepts of Generic Paths.

The *NetInf Information Network Interface* (INI) at the right is the collection of NetInf protocols which are used to communicate to other NetInf nodes in the network. There are three levels of protocols corresponding to the additional services, name resolution and transport control levels just described. In the middle we find the NetInf name resolution protocols. There are different protocols for different purposes, for instance, one protocol can be based on DHT technology, while another can be based on local search using IP multicast. Several protocols can coexist and used simultaneously. On top we find protocols for the additional services. At this level, there can be standardized protocols as well as proprietary protocols. Below we find the NetInf transport control protocol able to connect in and to wrap Generic Path mechanisms.

The NetInf application programming interface is a programming interface with functions that can be called by application programs, including NetInf additional services, in order to make use of the basic NetInf service. The NetInf API provides a communication service which borrows some ideas from the publish/subscribe communication paradigm. There are however no restrictions on events handling and asynchronous event notification.

The functions of the API are divided in two classes: functions used by entities making information available in the network–publisher functions, and functions used by entities retrieving information from the network–subscriber functions. The description also has a third class of functions that deals with the actual data transfer. This class is not part of the API proper, but rather functions provided by particular transport protocols.

Based on the requirements identified on the early phases of 4WARD concept the Information Objects infrastructure was built on some reference ideas. First one was the observed semantic overlapping of the name spaces on the internet and the way how information is delivered and consumed. Practically the Internet infrastructure is forced to make the major leap from data delivery to information delivery. This means a transparent usage in terms of location and a meaningful approach in term of content value and this is a must in the ocean of bytes, users and connection possibilities.

The architecture is built around the Information Object concept. This approach aim to satisfy a demand from a consumer point of view, who asks for one piece of information, characterized by certain features, not for content placed at a location on the internet, in multiple forms and without any connections to other ones in the cyberspace.

The fundamental innovation of the Information Object is that it provides a means for directly referring to a piece of information, without the need for overloading locators and putting them in the role of being an identifier and a locator at the same time. Today, typical requests for a web resource made, e.g., with the well known Uniform Resource Locators (URL, e.g. http://www.4ward-project.eu/index.html) contain the host name or network address of the physical machine holding the information. Users are however rarely interested in the machine serving the request, but in the information itself—in our example the 4WARD webpage—and rather use the URL as an opaque identifier for a piece of information. Information Objects provide a location-neutral means of referring to these pieces of information, so that a user can directly request what he is actually interested in, the information.

10.4 Nuts and Bolts

The Information Model was designed as the fundamental piece for an Information centric architecture. The fundamental step forward is to offer improved meanings over the internet delivery of content and services. Due to the booming profile of mobile access, the current architecture doesn't take advantage of some specific aspects like: best location to deliver a certain copy of an information piece or how to take advantage about specific features of the content (e.g. for a movie, what kind of action contains, actors, etc. things known as data about data or metadata). We further distinguish between Information Objects just at metadata level (IO) and Information Objects as a reference to the actual bit patterns holding the payload. The bit patterns are referred to as bit-level objects (BOs). Examples for a BO are the well-known file and stream format such as an mp3 file, a video stream or a voice conversation. In this sense, the second usage of Information Objects can be regarded as the representation of a BO in the NetInf name resolution system. This two kinds of Information Objects have the key difference that second one are a reference to concrete bit-level objects, while first ones do not have a payload associated. IOs represent a higher-level, semantically meaningful entity like a certain song. As illustrated in the lowest level of Fig. 10.3, data objects may be stored at different locations in the network, possibly at a server hosting a file or serving a stream, but also at mirror servers, caches or user devices that have already retrieved this file and make it available to others. The common notion for referring to all these cases contains means to verify that retrieved data is authentic, such as a hash that is part of or securely bound to the name of the IO. A IO can be accessed, e.g., using a Generic Path (presented also in this book) that is optimized for NetInf transport.

Higher semantic levels can be expressed through IOs. In Fig. 10.3, the IO represents a certain song *Song1*, which is available in different encodings, in our example an mp3 version *Song1.mp3* and a wav encoding *Song1.wav*, each of which is

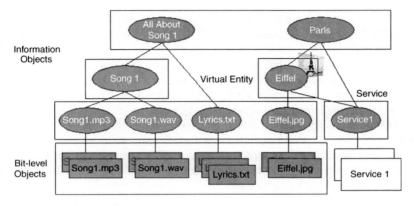

Fig. 10.3   A conceptual Information model

available at two locations in the network (indicated by the boxes on the BO level). From a user perspective, this means that it becomes possible to e.g. request a certain song without specifying a particular file, which significantly increases the number of suitable data objects. Again, this reflects the idea of adapting to the information consumption behavior of users: As long as the quality is sufficient and his device can play a certain encoding, the selection of the BO might be transparent for the user, he don't care about the actual recording of the song he wants to listen to. The objects are optionally encrypted so that classical Digital Rights Management methods can be applied, but also user-generated content can be protected against unauthorized use. IOs may be linked to other IOs which represent even higher levels of aggregation.

Is obviously known the danger to generate endless loops of bindings when we go accidentally or not to connect IO-s to BO-s used as IO-s. Therefore, the publishing mechanism must prevent this type of accidents using, by example, template schemes for IO generation and storage verification. As illustrated in Fig. 10.3, the *All about Song1* IO groups different pieces of information belonging to this information object. Apart from the song itself, this can also be objects of a completely different nature, such as a document containing the lyrics (see *Lyrics.txt* in the figure) or any other type of related information. Another example for these higher level semantics is a symphony, e.g., Beethoven's 9th. Different orchestras may have performed it, and even from one orchestra there might be several recordings of one and the same symphony, possibly in a lot of different qualities and formats. The IO provides a means to group all these under one common object that refers to the symphony in a very general sense. This is a particularly useful trait of the information-centric paradigm, as it enables the user to quickly access a piece of information, and deal with the different possible incarnations at a later point in time. If, e.g., a user is interested in Beethoven's 9th, but does not care about the performing orchestra or the particular audio encoding, the IO is a very powerful concept of expressing this request toward the Network of Information.

For such a system to work, establishing and maintaining the bindings between the IOs and BOs is crucial. Depending on the use case, different methods are conceivable. Bindings can be centrally managed by the owner or another entity appointed by the owner. In this case, changes to the bindings, e.g. updating the respective entry in the name resolution system, can only be done if authorized by this entity. In other cases, a community might maintain the links, similar to the way the Wikipedia online encyclopaedia is managed today. Also, if a trans-coding service has generated a new encoding of an object, it would be also be natural to automatically establish a binding to the source object. Finally, self-management mechanisms can analyze the metadata associated to objects and automatically create the links. This process is related to the searching process; the advantage with the approach described here is that bindings are pre-computed and don't need to be created before being able to resolve an IO, while the main disadvantage is the lower degree of flexibility and interactivity. In both cases, the work on ontologies that has been done in the context of the Semantic Web [7] can be very helpful. Also combinations are conceivable, e.g. a set of bindings is proposed by an automated tool and then reviewed and refined by an editorial team or community.

Updating bindings is also one approach to manage updates to information objects. An IO can for instance represent the metadata associated to the current issues of a newspaper, and when a new issue is released the binding from this IO to the respective BO is updated. Due to the fact that a BO can be interpreted and used in different scenarios, is possible to have associated IO's for each usage scenario, each update in a scenarion usage leading to the updated of the IO and associated binding.

The concept of the IO embraces a very wide range of entities conceivable as information, such as live streams, the sensor data or a communication session. Figure 10.3 shows the example of including the virtual representation of real world objects. In today's Internet, a typical way of retrieving such information is accessing a corresponding web page (possibly found by means of search engine), e.g., *www.tour-eiffel.fr* when you want to learn about the Eiffel tower. Recent proposals for an Augmented Internet [8] describe an automated mapping between physical objects and its virtual representation. Different mapping mechanisms can be employed, e.g. a combination of GPS location and image recognition software to find the virtual representation of a monument, or a Radio Frequency Identification tag to bind to a library book. In all cases, the corresponding virtual entity is an IO in the Network of Information, which can then be linked to all kinds of information. This information may be as simple as a picture of a monument (see *Eiffel.jpg* in the figure), but also all kinds of other data around this real-world object, such opening times, historical information or a ticket reservation service. A related field of application that is particularly useful for the operation of the NetInf system as such is the mapping of user identities into the Network of Information, e.g. from an ID card, Subscriber Identity Module (SIM) or government-issued signature card to a digital identity which can be used in the Network of Information, using of course specific adapters, since NetInf don't support directly these technologies. With such a binding, it then e.g. becomes possible to distribute access keys for Information Objects that can only be decrypted by a certain user or group of users. Also for accounting purposes, establishing a secure binding to the actual user is necessary.

As illustrated above, the NetInf object model can also represent services: A *Service1* can be available at different locations in the network. If different instances of a service are identical (e.g., different deployments of the same software package), they share a common ID on the IO level, while the individual instances can conveniently be located using the standard NetInf name resolution mechanisms (shown at Bit level objects). If, e.g., for load-sharing reasons, new instances of the service are created they can easily be added to the system under the same ID. Related, but not completely identical services can be grouped by means of an IO. We envision integrating a wide array of services, ranging from very tightly integrated services such as the storage service to very loosely coupled services that do not need to interact with the NetInf machinery.

All objects discussed above support associated metadata. This is crucial because the IDs of the objects are as such not meaningful to humans. The term metadata in this context refers to the whole of the information, other than the payload, associated with the object. Metadata e.g. consists of attributes like the bit rate and codec of an audio recording or the author and abstract of a document. Virtual representations of real-world objects might contain GPS coordinates, while an object representing a service will likely contain a detailed description of the features the service offers and who is responsible for the service. For objects relating to, e.g., a voice conversation, information of the parties involved in the communication will likely be included. The meaning of the metadata can differ depending on what kind of classification and rules have been specified for the metadata, this relates to work done on ontologies in the semantic web. Different ontologies attached to same parameters will render different interpretations (like street address can mean home, office or other address). The agreement on what specific ontology to use or how different ontologies should interoperate between different operators and users will help in supporting multiple resolution systems and bridge the challenge of different meanings of metadata in different contexts. The representation of metadata is covered in more detail in the information model which is presented in the next section.

The NetInf information model has two basic objects; the Information Object (IO) and the Bit-level Object (BO).

- The NetInf Bit-level Object (BO) is the basic data object itself, or the digital representation of the object if the object is not a digital object (this is the case for real world objects for example). In other words, the BO stores the data information of the object. NetInf treats this as opaque data and does not derive any semantics from it.
- The NetInf Information Object (IO) consists of three parts; the ID that gives the IO its name; the Metadata which contains the semantic information associated with object; and the BO, as described above.

In addition to naming the IO, the ID has certain security properties that enable NetInf to provide IOs that can be self-certifying, provide ownership information, etc.

The Metadata field consists of a set of attributes and provides semantic information about the IO. Metadata are used for security purposes and for search. Metadata are also of value for applications to understand how an IO can be used.

Fig. 10.4 Information model
IO format

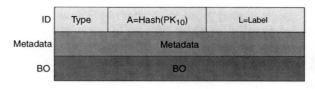

The data field can contain, by example, a BO handle, which is a reference to where the object can be found, or it can contain an IO ID, which is an indirection to another IO. It can be noted that what is stored in the data field of one IO can be the metadata of another IO, i.e., a set of metadata can be stored as an IO.

In today's host-centric networks, trust in information is based on trust in the infrastructure, including the hosts providing the data, the communication channels, and the Name Resolution Service (NRS) (e.g., Domain Name System (DNS)). In NetInf, due to the adopted naming scheme with built-in security properties, trust in information is based on the information itself, which is better suited to inherent needs of information-centric networks. The proposed naming scheme supports a combination of security-related properties including name persistence, self-certification, owner authentication, owner identification, and name certification all at once, going beyond the existing naming schemes for information-centric network architectures. The properties are ensured by the naming scheme along with security metadata associated with the information to be protected. Moreover, the NetInf naming scheme also provides flexibility and extensibility by supporting multiple types of data identifiers.

Pieces of information in NetInf are called Information Objects (IOs). Each IO is given a globally unique identifier (ID) according to the identifier/locator split paradigm, thus ensuring that the IDs are not bound to location. This location independence of IO's IDs enables multiple copies of the same (or similar) information content to be simultaneously stored under the same name at different locations, which itself results in a more efficient, large-scale data dissemination. Consequently, name persistence with respect to location changes is the fundamental property of the NetInf naming scheme, which is also satisfied by other information-centric network architectures such as DONA [9], PSIRP [10], and CCN [11].

In addition to location changes, the NetInf naming scheme also ensures name persistence of IOs with respect to changes of content of dynamic IOs, which means that the IO's ID can remain the same even if the data content changes. Of course, this property by itself is easy to achieve, but is a challenge in combination with another fundamental property offered by the NetInf naming scheme, that is, self-certification. Self-certification ensures that the integrity of data content of an IO can be verified if its ID is authentic. In other words, this property means that any unauthorized change of data with a given ID is detectable.

Self-certification of static content can be achieved by simply including the hash of content in the ID, but, for dynamic content, this would violate name persistence. Self-certification of dynamic content can be achieved by signing the hash of the self-certified data by a secret key corresponding to an adopted digital signature algorithm, by including this signature in the associated metadata, and by including the public key corresponding to the used secret key in the ID. This public key is then used for verifying the signature. More precisely, only a shorter, hashed value of the public key needs to be made part of the ID, whereas the public key itself can be put in the security metadata. Therefore, if the signed content changes, then the signed hash of this content changes, but not the ID and, yet, the data integrity can be verified by using the public key corresponding to the persistent ID. This guarantees that the retrieved content is authentic if its ID is authentic, provided that the signature is verified as valid. Since the ID can also contain other information, which also needs to be authenticated, the ID as a whole should be included in the self-certified data. Authentic ID retrieval can be achieved by using recommendations, past experience, and specialized ID certification services and mechanisms.

Accordingly, satisfying both self-certification and name persistence implies that the NetInf IDs need to be flat, at least partially. This means that a hierarchical NRS may not be sufficient for the name resolution and that other solutions should be used instead, e.g., a solution based on multiple distributed hash tables. In turn, flat IDs are advantageous with respect to mobility and can be allocated without an administrative authority by relying on statistical uniqueness in a large namespace.

The basic ID structure of an IO in NetInf is now explained in more detail. In view of Fig. 10.4, an IO is formally defined as IO = (ID, Data, Metadata). Data contains the main information content of the IO. Metadata contains the information needed for the security functions together with any attributes associated with the IO, e.g., describing the audio or video content contained in the Data or the owner of IO. Alternatively, Metadata can be stored independently, as a separate IO. An owner of an IO is then defined as any entity able of creating or modifying the IO. The IO's ID is formally defined as ID = (Type, A=Hash(PK), L), where A=Hash(PK) is the Authenticator field containing the hash of the public key PK associated with the IO, L is the Label field containing arbitrary identifier attributes, and the standardized Type field, for flexibility and extensibility, specifies a particular type of ID, e.g., the hash function(s) used to generate the ID and the variable format and structure of the label and how to interpret this structure. In particular, for static IOs, the hash value of the content is included in the label. Any entity knowing the secret key SK corresponding to PK from the IO's ID can thus be regarded as an owner of an IO.

It is important to note that, unlike other proposals like DONA, the authenticator field corresponds directly to the IO and not to a physical entity controlling the IO (e.g., an owner). This enables another distinctive property of the NetInf naming scheme, i.e., name persistence with respect to changes of owner or owner's organizational structure. This property of owner independence can be achieved in two ways, by the less complex basic approach and the more complex and more secure advanced approach. In the basic approach, the PK/SK pair from the IO's ID is securely passed from the previous owner to the new owner, whereas in the advanced approach, the previous owner authorizes a new PK/SK pair to be used by the new owner by an authorization public-key certificate. In both approaches, a previous owner of an IO is trusted to authorize a new owner of the IO. In the advanced approach, an owner of an IO is any entity knowing the secret key SK or any other secret key authorized by SK.

Both the approaches technically allow all legitimate owners in the certificate chain to make valid changes to the IO. If this behavior is undesired and former owners should be prevented from making changes to the IO, then the advanced approach facilitates prohibition on a legal basis, by including the production and expiry times in each authorization certificate and by providing a trusted time certification service to the involved owners. Alternatively, a key revocation mechanism can be used for this purpose as well as for the revocation of compromised secret keys. Note that an information-centric network is well-suited to key revocation, because the key revocation lists can be published as the associated IOs.

Trust and accountability in NetInf is achieved by two mechanisms, namely, owner authentication and owner identification. With respect to owner authentication, the owner is recognized as the same entity, who repeatedly acts as the object owner by demonstrating knowledge of the same owner's secret key, but may remain anonymous [12]. With respect to owner identification, the owner is in addition identified in terms of a real-life identifier, such as a personal name. This separation is important to allow for anonymous publication of content, e.g., to support free speech, while at the same time allowing building up trust in a potentially anonymous owner as an entity in possession of the same owner's secret key. Yet another distinctive feature of the NetInf naming scheme is that owner authentication is separated from data self-certification, by allowing the PK/SK pair used for owner authentication to be different from the one used for data self-certification.

Owner authentication is proposed to be achieved by including the hashed public key of the owner in self-certified data and by signing this data both by the self-certification secret key and the owner's secret key. Owner identification can then be achieved by including both the owner's hashed public key and the owner's real-life identifier in self-certified data and by signing this data in the same way as for owner authentication. In addition, the owner's real-life identifier also needs to be verified, and this can be achieved by using and verifying an additional signature, i.e., the public-key certificate binding the owner's public key to the owner's real-life identifier. This certificate is issued by a trusted third party upon verifying that a physical entity knows the owner's secret key and has the identity attributes specified by the owner's identifier. All the needed signatures are included in the security metadata [12].

10.5 Operation

Information centric approach needs a serious revision of the current communication architecture. Due to the need of new functionalities the access nodes will expose capabilities higher that e.g. TCP/IP transports level. The need to have in one place the functionalities specific for Object level delivery modifies accordingly the way how applications address the network. In the following is presented the structure of the services provided by NetInf functionality and some scenarios for usage.

From a network architecture perspective, NetInf eco-system consists of a number of components; a resolution service, a storage service, client applications and a number of NetInf additional services, see Fig. 10.5.

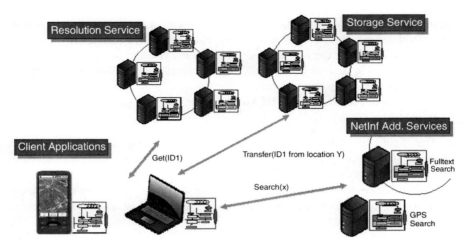

Fig. 10.5   NetInf network architecture

In NetInf a network node and a terminal has the same node architecture. A device that in a traditional network would be regarded as a terminal/end-point will in the information-centric NetInf network be a network node interconnecting information and real world objects.

Before getting into details let us give an example of how the components of a NetInf eco-system can interact, in this case an Augmented Internet scenario.

As mobile Internet-enabled devices become more ubiquitous, users will want to use them to access greater amounts of information and services in the real world. This includes objects close to the users such as everyday objects, people they meet, or places they visit. When accessing information on the move, it is essential that the accessing information does not detract the user's attention from the real-world activities. Unfortunately, mobile Internet access as we experience it today requires a lot of attention by the user and is therefore not suitable for many scenarios. To support such scenarios, a smooth integration with the real world that enables service access without interrupting the user's real-world work flow is needed by Internet applications. However, such applications are difficult to build on a large scale due to the current Internet architecture. Basically it does not provide a notion for real-world integration.

An Augmented Internet paradigm, for example, when a tourist near the Eiffel Tower cares about opening hours, ticket cost, the history of the monument, and so on. Whether this information is located on a server in Paris or elsewhere is irrelevant to the user. URIs such as www.tour-eiffel.fr provide an abstraction layer to the location of information, but nevertheless tie it to specific network nodes. Such applications pose two main requirements on an underlying infrastructure. First, the Augmented Internet needs a notion of virtual representations for physical entities that can cumulate and provide access to physical-entity-related services. Second, the Augmented Internet has to build and maintain a binding between the physical entity and its virtual representation on the future Internet. The NetInf addresses these requirements by providing an API that is common for objects that represent real-world entities as well as objects that represent services, content, and other digital entities. Based on the common API and the information model, bindings and interactions can be defined between the objects representing real-world and digital entities to enable Augmented Internet applications.

If the user in our Augmented Internet use-case wants to get the menu from the pizzeria closest to the Eiffel tower the following will happen:

1. The client node will compose a request for an IO with a set of attributes (e.g. object type = menu, restaurant type = pizzeria, location of restaurant = closest to the Eiffel tower, my current location = ?GPS).
2. The resolution service (NRS) resolves on the attributes and identifies matching IO(s); these are returned to the client node. This is typical a way to implement searching functions.
3. The client node selects the IO(s) it wants to retrieve. If the IO contains a BO handle, the BO is requested through the appropriate transport interface (including local storage or cache). If the IO contains a reference to another IO, recursive IO resolution can be done, until a final BO is found and selected by the user.
4. Finally, the client node requests the corresponding BO(s) to be transferred by currently available transport mechanisms.

NetInf allows a two-step resolution process. In the first step, an ID or set of attributes is resolved to a set of IOs. The application/user then selects one or more IOs for retrieval of the corresponding BO. The reason for this is to give the application/user control of which IOs are retrieved. Retrieving different IOs might incur different costs; they might be available at different download speeds, etc.

To optimize NetInf performance, when latency is a priority, NetInf could assist the user/application with this selection, thus hiding the second resolution step. The selection could then be made in conjunction with the first resolution step avoiding an extra round trip delay. Network based selection can also relieve the end-system/user/application of the need to choose among BOs to identify the "best" one; this selection can be done by the name resolution system directly. As a consequence, NetInf could mimic user-level behavior as known from client/server protocols (like HTTP) without sacrificing flexibility. If such a NetInf-executed selection process should take user preferences into account there would be a need to add an API where the user can set policies and preferences for such a selection process (else, default policies would be applied). This has so far not been addressed in our current work.

10.6 Evolution

NetInf aims to show relevant improvements versus current Internet practices. Therefore, the new architectures and processes need some methods to reveal the progress, the evolution. Here, the strategies used on this purpose are the feasibility estimations, simulations and the prototypes. One key component on the Information Centric architecture is the Name Resolution system, as described before, and his performance is crucial for successful usage of the concept in real life implementations. An integrated name resolution (NR) and routing scheme is one in which the network performs both NR and the routing of the request to the destination in a single step, as part of a unique process. As the object retrieval path is concerned, the transfer of data can use the topological shortest path to the requester or follow the same path of the request backward. In the latter case, caching strategies can be implemented by the intermediate nodes. It is worthwhile to notice, however, that when an integrated routing architecture is adopted, even if the optimal shortest path is used for the data transport, the request has to be routed through the NR nodes, and then the overall resulting routing path can be sub-optimal when compared to a pure topological routing policy. Therefore, an interesting question to be answered is what the resulting stretch for an integrated NR and routing policy could be. The routing stretch is a measure of the effectiveness of a routing algorithm with respect to the optimal shortest path routing. Therefore, it can be defined as the ratio of the length of the actual routing path and that of the shortest path.

The Multiple DHT (MDHT) architecture is one possible approach for implementing an Integrated Name Resolution and Routing System in a Network of Information. MDHT runs over a distributed Dictionary, that is, a data structure for name resolution distributed over the infrastructure of NetInf nodes which contains binding records for name resolution. In order to estimate the scalability of MDHT, we have chosen a scenario where, for simplicity, a global Dictionary is implemented with 4 hierarchical DHT levels. A simple analysis shows that, with current available storage technology (e.g. Tera-RamSan), up to $O(10^{15})$ binding records of 1 KB each, on average, can be managed over the global Internet. In that case, a Dictionary of 4 TB Storage with $O(10^9)$ binding records is required on each NetInf Node of the Global Internet, assuming about $O(10^6)$ NetInf Nodes in the Internet. This estimation takes in account just how the multiplicity of some flat DHT levels is reflected in a multi-level approach.

With a rate of 2 requests per second per user, a NetInf node can then handle around 8300 users with a single Tera-RamSan Storage Unit. By distributing the same Dictionary partition over multiple storage units on the same node and by parallelizing, it is possible to further increase the request rate per user or the number of users per node. These figures are, of course, considered just for local requests. Due to the multi-layered nature of MDHT and the way how changes are propagated, we think is reasonable for the current analysis.

The latency of a single resolution operation can be estimated to be less than 500 μs. Indeed, if we assume $O(10^9)$ binding records per node to be stored in a Balanced Binary Tree data structure on the same node, around 30 Tree Depth Levels are needed. Since each access operation requires about 15 μs, each resolution request can be processed in about 450 μs.

As to the refresh of the binding records in the Dictionary, this process must have a frequency that generates an acceptable control traffic load on NetInf nodes.

A "change slowly, react quickly" strategy allows keeping low the frequency of re-fresh packets. This strategy permits the network to change its state slowly, but to react quickly to the actual user requests. Under this assumption, a complete round of Dictionary refreshes lasts about 5.6 days with a required bandwidth of 10 Mbps per NetInf node handling 10^9 objects.

The MDHT feasibility analysis points to performance levels that appear to be quite good when compared with analogous results from the DONA proposal. Note that, presumably, also other DHT-based approaches show comparable results with those of MDHT, since the above considerations can easily be extended to a generic DHT-based approach.

Another referential situation is the one characteristic for Cooperative Multiac-cess. Multiaccess for wireless networks is a very active research field but, until now, previous work focused only on integrating wireless LANs with cellular networks. In the near future, with mobile WiMAX deployments increasing, many expect fierce competition between 3GPP and WiMAX Forum technologies for delivering mobile broadband services. We take a different approach in this work and study cooperative multiaccess in a large-scale urban scenario, where mobile WiMAX and 3G cellular are used in a cooperative manner to deliver mobile video services. We position our work with respect the novel information-centric approach taken by NetInf and eval-uate through simulation the benefits arising from better management of multiaccess device capabilities in wireless metropolitan area networks (WMAN) multimedia content distribution.

The simulation scenario discussed in this section we consider N connected and active NetInf nodes moving in a rectangular metropolitan area of 7.5 km^2 (5×1.5 km). By "active" we mean non-stationary nodes that request broadband network services. Nodes that are in (multiaccess) paging mode only are not con-sidered in the simulation as they do not consume network resources suitable for broadband services. In each scenario, 20% of N represents pedestrians, 60% users in automobiles, and 20% passengers on a train. Cars and pedestrians move accord-ing to the random direction mobility model with velocities set to $v_a = 50$ km/h and $v_p = 0.5$ m/s, respectively. The train moves at a constant linear velocity $v_t = 60$ km/h. Train passengers are randomly distributed in 6 train cars; the total length of the train is 200 m. In the multiaccess scenarios, all nodes can connect to both the mobile WiMAX (R1) and the 3GPP (R2) cellular networks, and support media-independent handovers using the abstraction mechanisms detailed in [13]. Further simulation details are given in [14].

The following scenario (Fig. 10.6) illustrates the case of two overlapping WMANs, which use two different technologies, which we will refer to as R1 and R2, respectively. Both types of WMANs can provide broadband streaming services to mobile nodes. In host-centric session mobility, the video flow always originates from a particular video server S (on the right side of Fig. 10.6) and terminates on the mobile node R (on the left side), passing through any of the networks in be-tween. Authentication and authorization would be based on geolocation and per-sonal account mechanisms. This is quite familiar to researchers using online doc-ument archive services such as the ACM Digital Library or IEEE Xplore. Today,

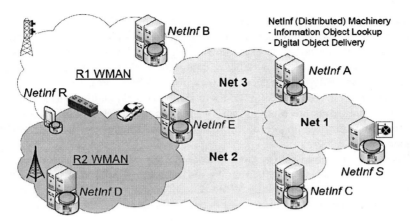

Fig. 10.6 A NetInf mobile multi-access scenario

the same laptop is authorized to access IEEE Xplore, for example, if connected via the institutional network but it is not authorized if it is connected through a public access network. In an information-centric network, the laptop should be able to access the database irrespective of its attachment point. More specifically, the point of attachment plays a lesser role in NetInf, as information takes center stage. Today, on the client (mobile node) side, a video stream in presumed to be legitimate because S appears to be trustworthy as per the URI of the video stream although it is well-established that URI spoofing is not uncommon.

In conventional host-centric mobility scenarios, S acts as the "correspondent" node (CN). Once a connection is established CN remains the same. However, in an Akamaized network (or for that matter a BitTorrent P2P network) aimed at information dissemination (such as YouTube-type video repositories), the correspondent node will be chosen based on end-to-end metrics and sophisticated algorithms based on DNS resolution. While the mobile node is connected to R1, the overlay CN selection algorithm may opt to connect the receiver with node B. In a host-centric paradigm, when the mobile node hands over to R2, the CN will remain the *same* (B or S), although this is not necessarily the best choice anymore. On the contrary, in the information-centric paradigm the video stream could originate from any of the NetInf nodes A through E and S, depending on end-to-end metrics *and* information from in-network management entities. The streaming object would be self-certifiable, and thus it is not mandatory to originate solely from S or its Akamaized copy at B) and the requesting NetInf node would be authenticated and authorized based on self-certifiable credentials, irrespective of its location or current point of attachment (R1 or R2).

This is a concise summary of the study of this NetInf-inspired cooperative WMAN access scenario. We simulate scenarios where (a) either of the WMAN technologies is used to deliver network connectivity and services independently and compare it with (b) the case where the two network technologies cooperate. First, all NetInf nodes, although inherently multiaccess capable, are configured to use only

R1 as they move in the observation area. In this case, mobile nodes are able to perform only intra-technology handovers. We repeat the simulations; using exactly the same paths for each node i in each run r configuring all nodes to use R2 only in a similar manner. Finally, it is possible to configure all nodes to dynamically choose between R1 and R2 as their access network in the entire metropolitan area. We can enforce policies that direct NetInf nodes to avoid vertical handovers as long as their service requirements are met, but aim at ensuring that each NetInf node receives its streaming information object throughout several changes of network point of attachment.

These type of simulation results indicate that an information-centric approach, based on self-certifying objects, may be instrumental in achieving significant performance improvement in future WMANs. Of course, our evaluation framework can be further enhanced with several detail levels. Besides, the scenarios considered only a subset of services of the future Internet. Nonetheless, the preliminary results are very promising and indicate that there is a lot to be gained by following a cooperative information-centric, rather than an antagonistic host-centric approach.

10.7 Conclusions

We have taken the information centric paradigm as the basis for our work. From this we have developed an information model encompassing not only virtual data objects but also real world objects as well as services. To make it possible to design a new information centric network architecture that is more scalable and has better security properties than today's Internet architecture one key component needed is a new naming scheme. Through the design of our naming and security framework we have been able to build an architecture that provides security for the information objects themselves rather than for the boxes containing them and the links interconnecting the boxes.

The information-centric communication abstraction has a number of advantages. Efficient distribution of content to a large set of recipients can be implemented. Content caching is built into the architecture and is thus provided without resorting to either interception of requests or special configuration at the receiver. Load-balancing is provided without depending on add-ons such as DNS round robin. Both reliability and performance are improved, since information can be retrieved from the closest available source.

Performance and reliability can be enhanced by an information-centric paradigm in a heterogeneous wireless environment where there are disruptions in communication, transient access opportunities and multiple access choices. An information-centric communication abstraction gives more flexibility in the delivery of data objects compared to using an end-to-end byte-stream. The network has better knowledge of the intent of the applications and therefore has the possibility to treat the data more intelligently. The network can easily deliver the data using multiple routes, redundancy over the available paths, and intermediate storage to overcome connectivity disruptions. The extreme is a scenario when an end-to-end path never exists.

With an end-to-end reliable byte-stream, the network has to violate the assumptions of the abstraction, or the application must implement these functions itself, including application gateways at suitable network locations.

The performance and reliability benefits from using storage also benefit non-dissemination applications, such as personal email. Another example is that direct delivery of email between two laptops with WiFi connectivity can be supported without involving infrastructure.

The information-centric approach gives new possibilities to prevent denial-of-service (DoS) attacks. With an information-centric approach nobody can force network traffic your way without your consent. This is made possible by moving control from the sender to the receiver. A sender can make information available, but for that information to be transferred, the receiver has to ask for it. Prevention of DoS attacks is a motivation to design NefInf as a replacement to the current TCP/IP for end-to-end communication [15, 16].

10.8 Related Work

Of course, those topics addressed inside NetInf are in the focus of more different initiatives in the networking research world. Here are some brief descriptions of some of the most relevant ones from our point of view.

In the area of information-centric networking, the Content Centric Networking (CCN) approach has many ideas in common with NetInf. One difference is that CCN uses hierarchical names. The hierarchies typically correspond to organizational structures. In CCN, the root of the name tree that constitutes the identifier for an information object has to be signed by an entity. The CCN security concepts require that this entity has to be trusted by the users. This means that when this organizational structure changes (e.g., an object changes owner or an employee changes organization), the object has to be republished under a different name.

Another closely related initiative is PSIRP. Their main idea is to implement a pure publish/subscribe information-centric system. They use rendezvous identifiers to retrieve the information objects. In addition, they have scope identifiers to restrict the distribution of objects. However, to our understanding, PSIRP does currently only focus on self-certification via the names but not on owner authentication and the other security properties supported by the NetInf naming framework.

During the previous sections we have detailed strong reasons why naming is important to NetInf. One is the use of self-certifying names: if you know that you have the correct name of an object, you can verify the authenticity of a received copy by applying some algorithm its contents. If the object is authentic, the algorithm will render the name you used to request it. This property is essential for NetInf as objects should be verifiable without having to trust the source from which the object is retrieved. Related work in this area includes Self-Certifying Public Keys and DONA.

The basic idea that a name contains an 'object owner'-related part that can be used for authentication and a 'label' part that is under the control of 'the owner' has been borrowed from DONA. One limitation of DONA is related to owner change: If the owner changes, the name also changes. In our naming scheme, we can keep the name persistent even when the owner changes. This is done by using a chain of certificates that is stored in metadata and securely bound to the name. One major criticism of DONA has otherwise been its poor scalability with regards to name resolution because of the use of flat names. Similar criticism could also be made toward our proposal. Especially as NetInf is designed to scale by the order of 10^{15} objects. To tackle those problems, we have investigated two approaches for scalable name resolution in NetInf: the Multiple DHT (MDHT) and Late Locator Construction (LLC). Related work in this area includes the Unmanaged Internet Protocol (UIP), ROFL, and i^3.

References

1. V. Jacobson, M. Mosko, D. Smetters, J.J. Garcia-Luna-Aceves, Content-centric networking, Whitepaper, Palo Alto Research Center (January 2007)
2. R. Moskowitz, P. Nikander, IETF RFC 4423, Host Identity Protocol (HIP) Architecture (May 2006)
3. I. Stoica, D. Adkins, S. Zhuang, S. Shenker, S. Surana, Internet indirection infrastructure, in *SIGCOMM'02: Proceedings of the 2002 Conference on Applications, Technologies, Architectures, and Protocols for Computer Communications* (ACM, New York, 2002), pp. 73–86
4. H. Balakrishnan, K. Lakshminarayanan, S. Ratnasamy, S. Shenker, I. Stoica, M. Walfish, A layered naming architecture for the internet, in *Proceedings of SIGCOMM'04*, Portland, Oregon, USA, 2004
5. B. Ahlgren, J. Arkko, L. Eggert, J. Rajahalme, A node identity internetworking architecture, in *Proceedings of the 9th IEEE Global Internet Symposium*, Barcelona, Spain, April 28–29, 2006, In conjunction with IEEE Infocom 2006
6. V.G. Cerf, S.C. Burleigh, R.C. Durst, K. Fall, A.J. Hooke, K.L. Scott, L. Torgerson, H.S. Weiss, Delay-tolerant networking architecture, RFC 4838, IETF (April 2007)
7. W3C, Semantic web activity, http://www.w3.org/2001/sw/
8. R. Want, K.P. Fishkin, A. Gujar, B.L. Harrison, Bridging physical and virtual worlds with electronic tags, in *Proc. SIGCHI Conf. on Human Factors in Computing Systems* (ACM Press, New York, 1999), pp. 370–377
9. http://psirp.org/
10. http://radlab.cs.berkeley.edu/wiki/DONA
11. V. Jacobson, D.K. Smetters, J.D. Thornton, M. Plass, N. Briggs, R.L. Braynard, Networking named content, in *Proc. 5th ACM International Conference on emerging Networking Experiments and Technologies (ACM CoNEXT)*, Rome, Italy, December 2009
12. C. Dannewitz, J. Golic, B. Ohlman, B. Ahlgren, Secure naming framework for information-centric networks, in *ACM CoNEXT 2009*, Rome, Italy
13. K. Pentikousis, R. Agüero, J. Gebert, J.A. Galache, O. Blume, P. Pääkkönen, The ambient networks heterogeneous access selection architecture, in *Proc. First Ambient Networks Workshop on Mobility, Multiaccess, and Network Management (M2NM)*, Sydney, Australia, October 2007, pp. 49–54
14. K. Pentikousis, F. Fitzek, O. Mammela, Cooperative multiaccess for wireless metropolitan area networks: An information-centric approach, in *Proc. of IEEE International Conference on Communications 2009 (ICC 2009)*, Dresden, Germany, June 2009
15. B. Ohlman, B. Ahlgren, M. Brunner, M. D'Ambrosio, C. Dannewitz, A. Eriksson, B. Grönvall, D. Horne, M. Marchisio, I. Marsh, S. Nechifor, K. Pentikousis, S. Randriamasy, R. Rembarz,

E. Renault, O. Strandberg, P. Talaba, J. Ubillos, V. Vercellone, D. Zeghlache, 4WARD Deliverable 6.1: First NetInf architecture description, FP7-ICT-2007-1-216041-4WARD/D-6.1, Technical report (January 2009)

16. M. D'Ambrosio, M. Marchisio, V. Vercellone, Authors: B. Ahlgren, M. D'Ambrosio, C. Dannewitz, A. Eriksson, J. Golić, B. Grönvall, D. Horne, A. Lindgren, O. Mämmelä, M. Marchisio, J. Mäkelä, S. Nechifor, B. Ohlman, K. Pentikousis, S. Randriamasy, T. Rautio, E. Renault, P. Seittenranta, O. Strandberg, B. Tarnauca, V. Vercellone, D. Zeghlache, 4WARD Deliverable 6.1: First NetInf architecture description, FP7-ICT-2007-1-216041-4WARD/D-6.2, Technical report (January 2010)

Chapter 11
Use Case—From Business Scenario to Network Architecture

Martin Johnsson and Anna Maria Biraghi

Abstract One describes how 4WARD processes, concepts and technologies can be used and applied as to provide a suitable network architecture to support a futuristic business scenario. One describes the actors and the network environment of a futuristic business scenario, the 'AdHoc Community', which is followed by an analysis and extraction of a set of non-technical business-related requirements. Those are further analyzed and then mapped onto a set of technical requirements. The Design Process is then followed in order to derive a suitable network architecture made up of components and interfaces, which can be deployed into a physical network infrastructure. At the end, a discussion on design options is provided, as well as a comparison with a solution based on existing technologies.

11.1 Background

11.1.1 Community-Based Networks

Until very recently, Internet users were only spectators, i.e., they only received the content that was being made available by few providers on central servers. A gradual change is occurring, since users are progressively becoming producers and suppliers of content, knowledge, connection, bandwidth, context, etc. Among the world-famous and most used Internet portals of these days are the ones where users contribute to the site: YouTube, Wikipedia, Facebook, Twitter are just some examples.

M. Johnsson (✉)
Ericsson Research, Stockholm, Sweden

A.M. Biraghi
Telecom Italia, Turin, Italy

L.M. Correia et al. (eds.), *Architecture and Design for the Future Internet*, 225
Signals and Communication Technology,
DOI 10.1007/978-90-481-9346-2_11, © Springer Science+Business Media B.V. 2011

Even news sites today offer the option for users to contribute by sending news, photos or videos (e.g. the newspaper 'El Pais' with their 'Yo periodista' (I, journalist) initiative). Most of the recent "special unscheduled" events have been shown to the world by nearby common people that work as "on-site reporters".

But new ideas can only be imagined and developed if the network can support them, e.g., social networks these days are always dependent on a portal, or similar. But why can't a user establish his/her own ad-hoc community "on the fly" with some friends or selected members for a given purpose?

The concept of an Ad-Hoc Community (AdHC) carries forward the community concept in a world, where no longer only long-lasting communities will deal with a larger set of items, but where additionally short term ('ad hoc') communities are created for very specific purposes. The Ad-Hoc Communities will have new peculiar characteristics and aspects that are not currently available in web communities, as specific and innovative features in the Future Internet can enable the creation of innovative communities according to 4WARD innovation.

Basically the 4WARD concept of information centric networks (NetInf) is a driver for such innovative communities. Major NetInf characteristics anticipated as drivers for Ad-Hoc Communities are highlighted below.

Today people have to search and navigate a lot when trying to find what they have in mind, and often they do not even know what they really are looking for. When at last they find something fitting their interest, they have to select the address of a server where the object resides. At this point, the site might own only an old version of the object (which might be no longer valid), the selected address might point to an altered or corrupted object, or the address might look correct but the accessed content might be maliciously changed. The 4WARD NetInf concept wants to solve these problems by an integrated information provisioning concept.

A problem quite common today is about broken links: it is a common experience that a bookmarked page that has been visited repeatedly is no longer available. This problem will be overrun by the naming persistence feature in NetInf. A search operation no longer will end up with 'file not found' or 'site not found' errors, but will point to the requested information object as long as the object is available somewhere in the network.

Assuring that each object name is unique, the naming feature in NetInf is also useful for security purposes. In fact it avoids malicious links, e.g., such as Disney's names used to 'drive' users to quite different sites. The naming uniqueness is defined to assure not only that an object is always reachable, but also that its content has not changed since the object was named.

Another significant difference from today is that the naming feature in NetInf assures that all instances of an information object have the same content, in particular there will not be two instances of an object with differing actuality. This avoids problems being quite common today like accessing accidentally an older version instead of the most up-to-date one. Today, it can be really difficult to decide whether the object found (e.g. the text of a law) is the final version or some interim one.

11.1.2 Business Models

11.1.2.1 Existing Models

When we talk about online communities, we think of well known, crowded communities growing around various topics that they can nurture over a period of time. The model starts from a topic that a group of persons is interested in: e.g., information sharing or sell, buy and exchange goods. Then the community is created, the value proposition is clearly stated and the native group of interest starts to work. If the value proposition is really clear and good, the native group starts to grow. The output of the model, and also its goal, is to build a base of people to target messages to: the larger this base is, the more valuable the community becomes, the higher its revenues grow.

11.1.2.2 A New Model to Support AdHoc Communities

Major Ad-Hoc Community (AdHC) characteristics will be:

1. AdHCs will answer needs or enable jobs
2. AdHCs will be co-workers with persons, doing jobs instead of today's workers, or helping them to work easier.

AdHCs Will Address Needs/Jobs   Let's take into account as examples two well known communities: Wikipedia and eBay. They were born as 'stand-alone' communities, aggregating around a 'single' topic. After winning success, both started to expand in several directions: Wikipedia gave birth to a lot of more specific WiKis covering different fields, found that the service was expensive and began to ask for donations from users. eBay acquired online payments company PayPal and the Internet communications service Skype. According to this well known paradigm, success can be metered by long lasting community life and by its wide expansion to include contiguous businesses.

In the new AdHC model this is not likely to happen any more, as AdHCs will be better used as new ways with which people will communicate to each other, share experiences, learn new abilities or jobs, console each other, disseminate news and relevant information, provide aid, and so on. In this sense, success of a provider enabling AdHCs will not be measured with the lifetime of the AdHCs but with the satisfaction of the users when using the AdHCs which can be measured, e.g., in the loyalty of a user to look for solutions within the AdHCs of a particular provider.

In the 4WARD paradigm any object of information (see Chap. 9) will possibly be able to set a new community. New AdHCs will probably be born from objects of information representing either persons or things and will provide a fast and secure way to share something in time-sensitive scenarios. Major needs that should to be covered by new AdHCs are:

• Trust: In everyday life, but above all in times of crisis, people will be looking for trusted sources for every reason

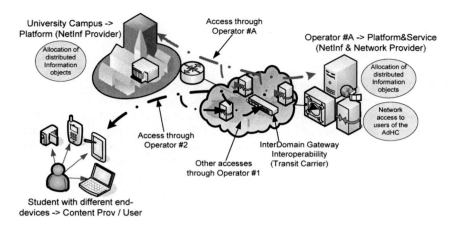

Fig. 11.1 AdHoc Community scenario overview

- Time: AdHCs deliver objects of information that are the most recent ones from the location which is closest to the recipient
- Security: AdHCs will only be a success if there will be easy-to-use and secure tools for submission, retrieval, and communication
- Traffic: the volume of traffic, depending on the number of contributors, will explode for impressive stories of large interest, little effort will be needed to develop the critical mass necessary for a community to be a valuable information source.

AdHCs Will Be Co-workers   As a co-worker, the AdHC will 'work' for people: it will be an enabler to improve confidence in the working group, despite the group members being in different locations and situations. It will also be an enabler in saving time to reach the proper objects of information and in being sure that they are trusted. It will work also to find the 'best fit' object from several possible points of view: e.g., only up-to-date contents, optimized presentation on the device in use, user interface optimized on user's skills, access to object location correlated to the traffic load and network characteristics. In this way, AdHCs will change also the process of community usage itself.

11.2 The AdHoc Community Business Scenario

11.2.1 Overview and Storyline

To show a short-lived community with a very specific objective, a use case was chosen with students setting up an AdHC to share resources/info/materials for the preparation of final exams at the University. Figure 11.1 provides an overview of the scenario.

Users may access the AdHC everywhere, through different connections and end-devices. Connections are provided by a Network Provider and access is provided by several alternative Access Providers. There is a Platform Provider that is responsible of providing NetInf, in terms of a basic cluster of services that reshape the information objects in a NetInf-ready fashion, thus enabling distribution, retrieval, (temporary) storing, etc. of the information objects. Network Provider A is the Operator that offers different access technologies. It has different interconnection agreements with the other actors.

The University Network represents the network that the students use to connect themselves to Internet from the campus. This network is connected to Internet thanks to the transit agreement between the University and the operators A, #1 and #2. One of the alternative Access Operators could be a Virtual Operator, as well.

In order to show different scenarios, the following interconnection models are assumed:

- Operator #1 has a peering agreement with the Operator #A. Therefore, there will be no charging based on traffic volume since it is assumed that these two operators will exchange traffic in a fair way.
- Operator #2 does not have any peering agreement with the Operator #A or #1. It can be assumed that this operator is in another country (e.g. students with an Erasmus scholarship in the other country use this network) and in the University it plays the role of Virtual Operator. It needs a transit, but it is not able to set up agreements as the exchange of traffic is in a very uncommon direction.

11.2.2 Roles, Actors, and Business Relations

The Business Model is based on the schema showed in Fig. 11.2. There will be three different roles for players:

- **NetInf Provider**—Group of University Network, Transit Carrier, AdHC communities already settled and Operator #A
- **Network Provider**—Group of Operators #1 and #2, Transit Carrier and Wireless Fidelity (WiFi) Operator
- **The User/Consumer**—Individual Users.

The NetInf Provider has a platform, that can be thought of as a library of basic functions, so that information can be easily prepared and delivered to users in a proper format. The platform will provide very ease-to-use functions and will be able to assist the user in all the creation of AdHCs, asking the user only for the most essential input. The NetInf Provider also provides the necessary network, so that each user can be connected through Operator A. Operator A will provide a platform so that the University Network and Operator A can interact. This information and data are stored (Arrow 1) in the Network Provider facilities (e.g. Data Centres), and the NetInf will pay it for this service (Arrow 4). This information will be organized in a simple and direct way, so that users can look for them directly (Arrow 5) and

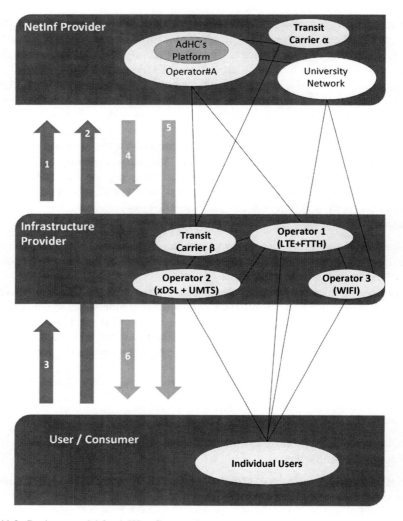

Fig. 11.2 Business model for **AdHoc** Community

access them by name. In case of metered information access. Users/Consumers will pay directly the NetInf Provider only for the specific information that are used from them. Otherwise users can subscribe to the Operator for a specific channel of information and the Operator will make the necessary agreements with (maybe several) NetInf Providers (see below, in the Network Provider role description). The Network Provider will provide the NetInf Provider with the necessary infrastructure underlying its services, allowing them to connect Users and Information Objects.

- Arrow 1—Network provider gives a Data Center to the NetInf
- Arrow 2—Payment of the use of information by the Users to the NetInf

- Arrow 3—Payment of Infrastructure access
- Arrow 4—NetInf pays for the storage to the Network Provider
- Arrow 5—Organization of the Information to be delivered to the user
- Arrow 6—Offer of infrastructure to access the NetInf.

The Network Provider role can be further split into Access Provider, that is required to access all destinations on the Internet, and Transit Carriers, that transport data among Access Providers. As can be seen in Fig. 11.2, Operator 1 (that gives access through Long Term Evolution (LTE)/Fibre To The Home (FTTH) technologies), Operator 2 (that gives access through Digital Subscriber Line (xDSL)/Universal Mobile Telecommunication System (UMTS) technologies) and the WiFi Operator are Access Providers, providing access to users through different technologies. As Network Virtualization will be enabled end-to-end, the Access Provider in the Future Internet architecture can be also a Virtual Network provider/operator: in this latter case, a cash flow will be set up from the Virtual Operator to the real Network Providers.

The Transit Carrier is needed when the Information Objects are not local in the Access Provider storage. In NetInf each object has a name. The naming addressing feature (see Chap. 5) can solve the name in local, when the object is located in the storage of the Service Provider (e.g. Operator x, University, Platform, Internet Service Provider (ISP), ...). Otherwise it will be released to a Tier1 Provider/Transit Carrier. This last player will then have the opportunity to cover a new role, as it will own NetInf capabilities that will enable it to solve names, authenticate the Information Object, recognize the domain to which the object belongs, forward the demand for that object in the proper domain and "link" the requested object to the requiring user (routing). Today the Transit Carriers traffic is already lowering due to localization: Peer-To-Peer (P2P) applications in fact are growing and reduce traffic transit exchange. Therefore Transit Carriers need to find new business opportunities. On one hand, Transit Carrier β can act as Access Provider, sharing business opportunities with other Operators. On the other hand, the Transit Carriers (e.g. Transit Carrier α) keep the service of "name solving" and a number of "core network capabilities", e.g. authentication, recognize the domain, forward the demand for a copy of the object in the proper domain, create a routing back, ... Therefore the Transit Carrier, or the whole group of Transit Carriers, become a kind of "root cloud" that allow global networking on the NetInf level and not only on the transport level as in the current networks: this service, maybe with different service levels, will be billed accordingly and a money flow will be established from the Operators to the Transit Carriers.

The User/Consumer will benefit from NetInf as follows:

- the naming uniqueness feature, that assures that accessed information objects are the original ones
- a proper protection to sensitive data is provided via the authentication feature
- the naming feature assures that the object is always delivered in its most recently updated version

- the naming persistence feature overruns the problem of broken links, as anywhere an object of information could be moved, its name is preserved and it will be always reachable.

The above listed characteristics correspond to the "core network capabilities" (see [1, Sect. 3.2.1]). These features provide higher quality access than in current internet or in P2P access, therefore the assured quality can be billed and a cash flow can be set between NetInf Provider and Operators/Access Providers.

The Access Provider could act also as the "one-stop-shop" that bills the users, then reversing to the other actors the charge according the agreements they have. So in principle users will pay the Network Providers both for access and information. The charging model will likely be a mix with a flat part related to access and a volume based part related to accessed contents and services.

The flat part will be shared among real and virtual Providers and the Transit Carriers, while the content related part will be possibly split among the content owner, the NetInf Provider (assuring security, quality and all the technical features) and the Network Providers (providing the required real or virtual infrastructure).

To win success in a business model, all the players must have a "winning perspective". In this proposed model, the following incentives can be seen:

- An incentive for all the Operators is to establish peering agreements, to keep costs down to zero if the traffic maintains the symmetry: this policy is good on very intensively used connection paths. If they exchange contents in order to reduce their traffic, they reduce also e.g. the number of ports dedicated to these inter-domain links, thus reducing also Capital Expenditure (CAPEX) for new ports and Operational Expenditure (OPEX) for the ports management.
- The NetInf paradigm for mobility provides the possibility to use "moving" storage (e.g. mobile phones or wireless connected pc storage) as temporary storage to save accessed Objects. Also this way NetInf enables Operators to reduce the physical infrastructure deployment.
- In case an Operator has not an agreement with the others, e.g. the Virtual Operator, it will have to pay on time or volume basis: NetInf enables to spend less on very unusual connections, taking advantage of the "moving" storage opportunities.
- The incentive for Transit Carriers, to contrast their losses due to volume charged traffic decrease, is that they will not get paid any more taking into account only the traffic volumes. The charging will be based on the "naming resolution" feature and on the different service level agreements they will have with the interconnected Operators: e.g. maintaining (long-lived) cashes for information retrieval and providing archiving capabilities.

11.3 Analysis of AdHC Business Scenario—Deriving Business-Related Requirements

Using the information provided in Sect. 11.2, it is possible to derive a set of user and business related requirements applicable to this scenario. These are summarized in Table 11.1. Each requirement has a 'slogan' followed by a 'description', and finally

Table 11.1 Business requirements

Slogan	Description	Comment
Usability	The scenario mandates an almost extreme level of ease-of-use, where the system shall be able to assist the user in all the creation of AdHCs, and with only the most essential input needed from the user	The easy-to-use requirement is due to overcome difficulties for low skilled users and to enable the AdHC set up by 'things'
Availability	The AdHC shall be available via all access networks and all devices, and generally be transparent as to what access network is in use. Contents will be always accessible (no broken links), despite how old they are	By default, the latest version will be retrieved. But if required, also older versions will be easily accessed by content/version tag
Timeliness	The AdHC service shall be provided without interruption, and with deterministic and generally short delays	Avoiding broken links and preserving names will allow any object of information to be always reachable
Security (user)	The identity of the user should have some degree of privacy options: visible to all, to some, to selected members, etc. in order to be unknown for users not in the 'white list' and to be protected against identity stealing	This as well as next 2 requirements are more general but important requirements, which have been added for completeness
Security (network)	The network shall allow only authorized users access to the network, and may shut off misbehaving users	
Security (content)	Content shall only be able to be accessed (including distribution) and/or manipulated by authorized users	Implicitly, this means that content is integrity-protected
Management (network)	The networks shall require only a minimum level of configuration and active monitoring in order to keep OPEX as low as possible	This is a critical requirement for future networks, see more details in Chap. 8
Charging Model (user–provider)	The charging model shall be built on the basis that the user only needs to have a contract with one provider or a broker. The contract shall support a separation of charges for general network access and for the services delivered (AdHC). The user should also be compensated if it provides capabilities that are useful for executing the AdHC service, for example cashing	

Table 11.1 (continued)

Slogan	Description	Comment
Charging Model (infrastructure)	The peering type of agreements among providers are preferred as to reduce traffic volumes. One-stop-shop to be preferred also in that it is easier for Users	
Charging Model (content)	Content can be itemized and consist of for example, storage, caching, and name resolution services	
Scalability (cost)	The cost for investing and maintaining networks are expected to stand at least in proportion to but preferably even better to number of users and the volume of exchanged traffic	This requirement follows up on the Management requirement above

a 'comment' field to provide additional information. The slogan which defines a type of requirements shall be thought of as being defined rather independent of the specific scenario outlined above.

It shall however then also be noted that the list of requirements shall in no way be viewed as exhaustive, but rather capture requirements essential for this scenario and the specific aspects it covers, but still comprehensive enough to deduce also the essential technical requirements which in turn can feed the design process with relevant input.

11.4 Refinement—Deriving the Technical Requirements

Besides the Business Requirements, it is possible to derive a set of technical requirements that enable the business scenario. These are summarized in Table 11.2. Each requirement has a 'slogan' followed by a 'description', and finally a 'comment' field to provide additional information. The slogan which defines a type of requirements shall be thought of as being defined rather independent of the specific scenario outlined above. The following list of requirements cannot be viewed as exhaustive, but is rather a list of essential technical requirements to support the described business scenario.

11.5 Applying the Design Process to Define a Suitable Network Architecture

11.5.1 Introduction

The Design Process is described in Chap. 4. In a future commercial setting, one should expect a highly formalized language and a sophisticated tool set to support

Table 11.2 Technical requirements

Slogan	Description	Comment
Session Continuity	The AdHC must be able to maintain a user-session, even if the user changes his end-device at a certain moment. This change of end-device could even imply the adjustment of the information object to a different format (same object and content, but several formats to be displayed), or even a switch in the access network for that client	This feature requires a previous agreement among stakeholders involved: for instance in the case of a change in the access network for a single client. Different end-devices might use different access networks to access the info objects (Service Level Agreement (SLA) implementation, information exchange during runtime)
QoS	The AdHC must support multiple content, so different QoS guarantees must be available. In particular: streaming must be supported (high BW) and real time (low delay and jitter) services must be available. If no guarantees are available, the user must be notified	This requires that the Ad-Hoc platform has access to the control systems of the network provider
Authentication & Authorization	The user must be authenticated and the content protected	The Ad-Hoc platform must provide Administration, Authorization, Authentication (AAA)
Integrity (network)	All the control mechanisms exposed by the network as services must preserve the network integrity	
Integrity & Privacy (contents)	The content that is uploaded/provided by a user must be preserved as it is (without any change in the content itself)	
Content Availability	End Users must have an unique Application Programming Interface (API) to request the different types of content, independently on the end-device they use to access the contents	
Traffic Engineering	Networks must have the capability to handle different traffic flows (point to multipoint, QoS characteristics, etc.) and adapt to important changes in the traffic matrix (e.g. offload of the high volume traffic flows to optical capabilities instead of its management at upper layers; the PCE (Path Computation Element) as a Governance Functionality able to work in a multi-layer environment)	This implies that the network provider must be able to handle traffic in an economically efficient way

Table 11.2 (continued)

Slogan	Description	Comment
Selection of the best content source	According to the end users demand, the NetInf must select the most suitable content source (which will take into account, e.g. data proximity, status of the content server, network congestion, type of user terminal) considering the end user experience	

and implement the Design Process. At this stage, and for this limited use case, we will apply the design process in a fairly simplified as well as informal way, though still be able to describe its generality and efficiency to go from the level of requirements to actual deployable network architecture.

The Design Process is supported by the Architecture Framework as described in Chap. 4, and more details about the Design Process can be found in [2] and [3]. The different constructs and entities making up this architecture framework (strata, netlets, components, as well as the design repository) will be used frequently in the following sections.

11.5.2 Requirements Analysis

In this phase the high level requirements are analyzed as to identify the network functionalities at the macroscopic level, i.e the strata. This analysis and its decomposition into strata is very much supported by the Design Repository. It provides high level design patterns that smoothly bridge from the technical requirements to a set of network functionalities. These design patterns consist both of reference strata [2] that define common and generic properties and functions that are used to build the actual strata, as well as a set of vertical and horizontal strata that can be tailored, via the reference strata, to provide the specific network functionalities needed to implement the technical requirements. Examples of reference strata could be to define a common signaling protocol as was described in [2] and in Chap. 4, but many others can easily be conceived. One example is the interoperability principles but many other examples could be found, such as security, QoS, mobility, name resolution mechanisms, generic routing protocols, policy-related functions, and not to forget the many different self-management algorithms.

In the following description we assume that design patterns defined as reference strata have been used in the process to identify and build the different vertical and horizontal strata. Thus, Fig. 11.3 shows how the technical requirements identify functionalities (logical nodes and the medium within strata) in the vertical and horizontal strata. It must here be understood that the strata may also contain other functionalities not described by the figure as the focus is here to see what specific

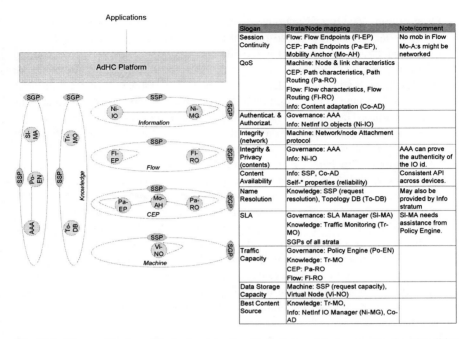

The table in the figure:

Slogan	Strata/Node mapping	Note/comment
Session Continuity	Flow: Flow Endpoints (FI-EP)	No mob in Flow
	CEP: Path Endpoints (Pa-EP), Mobility Anchor (Mo-AH)	Mo-A:s might be networked
QoS	Machine: Node & link characteristics	
	CEP: Path characteristics, Path Routing (Pa-RO)	
	Flow: Flow characteristics, Flow Routing (FI-RO)	
	Info: Content adaptation (Co-AD)	
Authenticat. & Authorizat.	Governance: AAA	
	Info: NetInf IO objects (Ni-IO)	
Integrity (network)	Machine: Network/node Attachment protocol	
Integrity & Privacy (contents)	Governance: AAA	AAA can prove the authenticity of the IO id.
	Info: Ni-IO	
Content Availability	Info: SSP, Co-AD	Consistent API across devices.
	Self-* properties (reliability)	
Name Resolution	Knowledge: SSP (request resolution), Topology DB (To-DB)	May also be provided by Info stratum
SLA	Governance: SLA Manager (SI-MA)	SI-MA needs assistance from Policy Engine.
	Knowledge: Traffic Monitoring (Tr-MO)	
	SGPs of all strata	
Traffic Capacity	Governance: Policy Engine (Po-EN)	
	Knowledge: Tr-MO	
	CEP: Pa-RO	
	Flow: FI-RO	
Data Storage Capacity	Machine: SSP (request capacity), Virtual Node (Vi-NO)	
Best Content Source	Knowledge: Tr-MO,	
	Info: NetInf IO Manager (Ni-MG), Co-AD	

Fig. 11.3 The specification of the horizontal and vertical strata and their respective functionalities and protocols

network features as addressed by AdHC use case the networks exhibit to support and provide the requested service.

There could be reasons to discuss whether the break down of the requirements into the chosen set of strata and logical nodes is appropriate. We have here followed a fairly traditional separation of functionalities and protocols that can be found in other networks and systems. Regarding the Information stratum, it is not really possible to find a background and history for how to deal with break down of functionalities, so for this stratum we rely on results both from 4WARD (Net-Inf) as well as Ambient Networks (Specifically their findings on Service-Aware Transport Overlays (SATOs) of which Content Adaptation makes up an integral part).

Thus the different strata are composed of as follows:

- Information stratum: The NetInf Information Object (Ni-IO) as described in Chap. 10, as well as the NetInf Manager (Ni-MG) which basically maps to Chap. 8. The NetInf Manager specifically aids in finding the best content source, and where different criteria can be used, e.g. "nearest" or "most compressed". Content Adaptation (Co-AD) is a piece of functionality that has been addressed in various research projects, for example Ambient Networks (SATOs... etc.). This stratum also contains protocols used for control and management of Information Objects and Content Adaptation functionality.

- Flow stratum: The Flow Endpoint (Fl-EP) contains functionality to terminate the traffic flows. Transmission and reception of contents are handled via the Flow Endpoints which puts contents into containers (e.g. data packets). Flow Routing (Fl-RO) is essential for the proper routing of traffic flows. Traffic flows can be established both with state in the networks as well as stateless. A certain traffic is dynamically bound to be carried by a Path provided by the Connected Endpoints (CEP) stratum. The protocols that are defined between Flow Endpoints are typically used for the transport of data packets, and then there is also a protocol for the management and control of routing of flows.
- CEP stratum: The Path Endpoint (Pa-EP) contains functionality to terminate Paths. Transmission and reception of data packets and handled via the Path Endpoints. Path Routing (Pa-RO) is essential for the proper routing of paths. The Mobility Anchor (Mo-AH) provides mobility support for paths, and which needs coordination with both Path Routing as well as the Path Endpoints. Paths can be established both with state in the networks as well as stateless. A certain path is dynamically bound to nodes and links in the Machine stratum. The protocols that are defined between Path Endpoints are typically used to carry information about the path (e.g. addresses of the Path Endpoints), and then there is also a protocol for the management and control of routing of paths, as well as a protocol for mobility management.
- Machine stratum: The Virtual Node (Vi-Node) represents the capabilities of a physical but virtualized node. Virtual Nodes are interconnected by physical but virtualized links. Across these links there are several different kinds of protocols: (a) one to control that only authorized Virtual Nodes are accepted into the Machine stratum, (b) one to carry data, and (c) for the management of the resources of the Machine stratum.
- Knowledge stratum: The Topology Database (To-DB) is the overall network database to store and cross-relate topology and resource status information that are registered by all the horizontal strata. In this way the Topology Database provides a generic approach and solution for name resolution. Traffic Monitoring (Tr-MO) keeps track of status of all paths and traffic flows, and which aids in traffic engineering, as well as in monitoring of SLAs.
- Governance stratum: AAA works similar to other AAA functions in existing networks, and which includes authorization for setting up new paths and flows. Unlike the case of existing networks, the Governance stratum has an enforcement role with respect to NetInf objects only to the extent its logical nodes have been set up to do so during the design process. Specifically, the nodes doing name resolution must prevent the registering of fake IOs. Paths and Flows that terminate in caches do not require access control or authorization unless the resulting IOs are registered within the Topology Database of the Knowledge stratum. The Policy Engine (Po-EN) is here primarily used to support the SLA Manager as well as making providing support for making decisions in regard of traffic engineering. The SLA Manager (Sl-MA) settles SLAs between networks dynamically in runtime.

11.5.3 Abstract Service Design

In this phase, we identify and compose the specific functionalities of strata and netlets. As we basically were able to get the strata defined in the previous section, we will focus here on the design and composition of the netlets. Each stratum is made up of a set of logical nodes and protocols, as well as the reference points (SSP and SGP). They were identified in the previous section, Generally speaking, we make each of them, logical nodes, protocols, SSP, and SGP a Functional Block (FB) that will constitute the fundamental building blocks for the creation of Netlets.

From [3], it follows that the separation of functionality should be so that functionality in horizontal strata goes into regular netlets, and functionality in vertical strata shall go into control netlets. However, it is also said in [2] that control functions of the network and which are not generally accessible by applications shall be part of control netlets. Some of the logical nodes in the horizontal strata are such control functions, and we will thus put also these in a control netlet.

For this example, we make a very simple classification of nodes:

- End systems which contain the applications, which in this example means the AdHC Graphic User Interface (GUI) + API, as well as the necessary transport capabilities of the horizontal strata.
- Network elements that basically act similar to current routers, with the addition to also manage information objects, including content adaptation.
- Network elements that provide management capabilities and which are an integral part of the network.
- Gateways that sit at the border of networks.

This results in that we should have four different types of netlets, where each of them contains a "package" of FBs that can be identified out from the strata definitions. Figure 11.4 gives an overview of the defined netlets.

- Data Transport Netlet: This netlet includes the FBs needed in a typical end system. This includes the NetInf Information Objects including also Information Management, as well as flow and connectivity endpoint functionality.
- Control Netlet: This netlet includes FBs needed for a typical router within a network, for example routing and mobility management capabilities. In addition, it also includes functionality for Information management and storage.
- Management Netlet: The FBs for the vertical strata.
- Gateway Netlet: All the SGPs of all the different strata.

In addition, the Data Transport Netlet and the Control Netlet also include the SSP of the Knowledge stratum, as there is a need for all functionality as part of this netlet to report as well as to retrieve status of topology, resource utilization, perform name resolution, etc. The Self* management property is an inherent capability of strata, and which is then also reflected when the strata functionality is loaded into netlets.

An additional observation is that the design of the Machine stratum will form input for the proper allocation of a VNet as being described in Sect. 4.6 above. The functionality within the netlets will use the Network Access component of the Node Architecture to utilize the resources of the VNet.

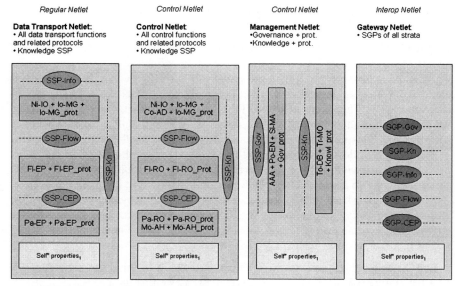

Regular Netlet   *Control Netlet*   *Control Netlet*   *Interop Netlet*

Data Transport Netlet:
• All data transport functions and related protocols
• Knowledge SSP

Control Netlet:
• All control functions and related protocols
• Knowledge SSP

Management Netlet:
•Governance + prot.
•Knowledge + prot.

Gateway Netlet:
• SGPs of all strata

1. The Self* properties is omni-present, and should also be part of the Management agent in the Node architecture.

Fig. 11.4 Netlet design

11.5.4 Component Design

In this phase, we select the specific software (SW) components that can be used to implement the different FBs in each of the netlets as was described in the previous section. For this example, and for the context of this book, we need to rely on components that are already available "off-the-shelf". In the case of not being available out from current standard solutions and protocols, we will instead point either the need for extensions to current standard solutions, or to the ongoing research activities within 4WARD. Table 11.3 provides an overview of the selection of components we have made for this example as to provide an implementation of each of the FBs. However, as solutions and protocols are being further developed that better matches the requirements for future networks, we should be able to replace the proposed components listed in the table below, to those developed for future networks.

11.6 The Deployable Network Architecture—Components and Interfaces

The deployable network architecture consists of a set of strata, where each of the Functional Blocks of the strata have been sorted as from node types into a set of netlets, and finally these FBs have been given a specific implementation by select-ing an appropriate component to realize the FB. The deployment of the network architecture is a highly automatic procedure thanks to the self-management proper-

Table 11.3 Selection of components for the implementation of each Functional Block for the Ad-Hoc Communities example

Functional Block (FB)	Component(s)	Comment/Reference
SGP-Info	See comment	This component corresponds to a specialized version of the INI being described as part of the NetInf node architecture, see Chap. 10
SGP-Flow	N/A	The Flow goes end-to-end
SGP-CEP	BGP	
SGP-Knowledge	See comment	A 'lite version' has been implemented in the CBA-based strata prototype, see Chap. 12
SGP-Governance	See comment	A 'lite version' has been implemented in the CBA-based strata prototype, see Chap. 12. A more elaborate implementation could be based on the Network Composition concept as developed in Ambient Networks (see [4])
Ni-IO + Io-MG + Io-MG_prot	See comment	The components correspond to entities being described as part of the NetInf node architecture, see Chap. 10, and specifically those related to the publication and storage of IOs
Ni-IO + Io-MG + Co-AD + Io-MG_prot	See comment	The components correspond to entities being described as part of the NetInf node architecture, see Chap. 10, and specifically those related to cashing and storage of IOs. Co-AD could be implemented as a SATO Port, see [4]
Fl-EP + Fl-EP_prot	TCP	
Fl-RO + Flo-RO_prot	N/A	TCP goes end-to-end
Pa-EP + Pa-EP_prot	IP	
Pa-RO + Pa-RO_prot	Intra-domain routing protocol, e.g. OSPF, IS-IS	

Table 11.3 (continued)

Functional Block (FB)	Component(s)	Comment/Reference
To-DB + Tr-MO + Knowledge_prot	See comment	To-DB could be thought of as an extended DNS with additional resource records, or alternatively for example implemented through an LDAP-based directory. A directory-based solution exists as part of the CBA-based strata prototype, and a DHT-based solution also exists as part of the NetInf prototype, see Chap. 12. Tr-MO can be implemented using GAP, see Chap. 8
AAA + Po-EN + Sl-MA + Governance_prot	See comment	Modules description in Chap. 2. AAA functionality can be defined using some DIAMETER-based solution, or potentially HSS/HLR Po_EN can be implemented using a policy framework, e.g., [5]
Self* properties	See comment	Self-organizing capabilities have been co-designed into the CBA-based strata prototype, see Chap. 12

ties of the strata and which become condensed into the self* properties FB in each of the netlets. For this simplified example, we can describe the procedure of self-deployment as a matter of dynamically finding out about the node characteristics of each (virtualized) node, and with this information at hand activate the proper set of netlets in each node of a network. Nothing hinders that nodes may come and go, i.e. this will be discovered by the self* properties FB and take proper action to re-arrange the topology of the network and then to install the proper netlets in new nodes.

Referring to Sect. 11.5 above, where the different networks and their roles were described, we can conclude that all the different networks would need all the different netlets except for the Data Transport Netlet, which is specific to the devices of the Individual Users. Here we also need to observe a slight over-simplification of the identified netlets, as the network elements of the networks that only plays the role of Network Provider don't need the functionality related to the Information stratum (i.e. Ni-IO, Ni-MG, and Ni-MG_prot). Thus, the netlet design should make two different versions of the Control Netlet.

One important observation that should be made here is that the Transit Carrier will play the role of a global name resolver, and this is made possible based on the fact that strata can compose themselves in accordance to a set of standard operations, see Chap. 4 above. In this use case, we can think of that the Knowledge strata of each network composes with each other, and that specifically the Knowledge stratum of the Transit Carrier performs a composition with Knowledge strata in other networks in accordance with the aggregation operation, resulting in that the Transit Carrier possesses knowledge of the locations of Information Objects throughout all the participating networks.

11.7 Conclusions

In this chapter we have described a business use case, the Ad-Hoc Community (AdHC), the roles emerging from the business case, possible actors and users. This formed input to define a set of business-level requirements and further on a set of technical requirements. The technical requirements in turn form input to the design process, where through a set of refinement steps which include the design of strata and netlets, as well as a selection of SW components, a deployable network architecture was derived.

We have shown that it is feasible to use the design process in order to derive a suitable network architecture out from a set of business-level and then further on technical requirements. The step-wise process of going from the macroscopic network-wide components known as strata, and then break them down on the microscopic level to Netlets and SW components, is straight-forward and easily conceivable to be implemented as a tool that could support a significant level of design atomization. Still, further research is needed to define a formalized support for the bridging of business level requirements with technical requirements in order to provide a fully seamless step-wise process integrating all the phases of the development cycle from business creation to deployment.

References

1. T.-R. Banniza, A.-M. Biraghi, L. Correia, T. Monath, M. Kind, J. Salo, D. Sebastiao, K. Wuenstel, Project-wide Evaluation of Business Use Cases, ICT-4WARD Project, Deliverable D-1.2 (2009-12-16) (not yet published)
2. M. Ángeles Callejo, M. Zitterbart (ed.), Draft Architectural Framework, ICT-4WARD Project, Deliverable D-2.2 (Apr. 2009), http://www.4ward-project.eu/index.php?s=Deliverables
3. M. Ángeles Callejo, M. Zitterbart (ed.), Mechanisms for Generic PathsArchitectural Framework: New release and first evaluation results, ICT-4WARD Project, Deliverable D-2.3.0 (Jan. 2010), http://www.4ward-project.eu/index.php?s=Deliverables
4. N. Niebert et al., *Ambient Networks: Co-operative Mobile Networking for the Wireless World* (Wiley, New York, June 2007)
5. A. Uszok et al., KAoS policy and domain services: Toward a description-logic approach to policy representation, deconfliction, and enforcement, in *Proceedings of Policy 2003*, Como, Italy

Chapter 12
Prototype Implementations

Denis Martin and Martina Zitterbart

Abstract An overview of the developed prototypes from which some are also publicly available from the 4WARD project web site is given. The prototype implementations of the architecture framework concepts are presented, showing the application and interactions of those concepts to other architectures described in the book: the virtualization architecture, and the In-Network-Management architecture. Since network virtualization also needs verification at a larger scale, parts of the concepts have been developed for a large virtualization testbed, which is also presented. The Generic Path concept for routing, forwarding and transport, and the Network of Information prototypes are described. In addition, their combination is outlined which shows that they nicely complement each other. Last, but not least, an integrated prototype showing the combination of Generic Path and In-Network Management concepts is shown, giving special focus on QoS aspects.

12.1 Introduction

To support the theoretical ideas outlined in the previous chapters of this book, some of them have been realized as prototypes. The experiences collected while implementing the different concepts gave valuable feedback and enhanced the ideas with crucial details. This chapter will give an overview of the developed prototypes from which some are also publicly available from the 4WARD project web site.

The prototype implementations of the architecture framework concepts described in Sect. 4.2 are presented in Sects. 12.2 and 12.4. These sections show the appliance and interactions of those concepts to other architectures described earlier in

D. Martin (✉) · M. Zitterbart
Karlsruhe Institute of Technology, Karlsruhe, Germany
e-mail: denis.martin@kit.edu

M. Zitterbart
e-mail: martina.zitterbart@kit.edu

L.M. Correia et al. (eds.), *Architecture and Design for the Future Internet*,
Signals and Communication Technology,
DOI 10.1007/978-90-481-9346-2_12, © Springer Science+Business Media B.V. 2011

the book: The virtualization architecture presented in Sect. 4.9 and the In-Network-Management architecture described in Chap. 8. Since network virtualization also needs verification at a larger scale, parts of the concepts have been developed for a large virtualization testbed described in Sect. 12.3. The Generic Path concept for routing, forwarding and transport, and the Network of Information prototypes are described in Sect. 12.5. In addition, their combination is outlined which shows that they nicely complement each other. Last, but not least, an integrated prototype showing the combination of Generic Path and In-Network Management concepts is described in Sect. 12.6, giving special focus on QoS aspects.

12.2 Designing, Running, Deploying Network Architectures

This section focuses on three parts of a network architecture life-cycle presented in Sect. 4.6: The tool-supported design of protocols (*Netlets*) using pre-existing building blocks, the test and evaluation of the designed Netlets with an open and flexible framework (*Node Architecture Daemon*), and the eventual deployment of the architecture into a virtual network using a virtual network (VNet) management environment. Those three parts can well stand on their own, but their close inter-working shows a key benefit of an integrated architecture design: The results of a development phase are seemlessly used as input to the next phases, and direct feedback (e.g. in the case of unexpected evaluation results) can be given to the previous phases.

12.2.1 Designing—Netlet Editor

In Chap. 4, we introduced concepts to support the design and execution of thousands of future networks and protocols. A first actual tool to support the network architect during the design process is the Netlet editor which exists as a prototype implementation.

The Netlet Editor utilizes models of existing protocol building blocks to support the design of Netlets. The description of the building blocks will be used later on to store information of building blocks in a repository and to implement dependency and constraint checking between building blocks. The tool will also ease the work of the network architect; however the architect still needs the domain knowledge (i.e. he needs to be a professional regarding network technology and design) in order to build proper protocols. We believe that completely automated composition approaches at run-time are not feasible due to the inherent complexity of protocol composition itself. Especially regarding important constraints of e.g. security properties, the limitations of fully automated composition approaches are not acceptable. Nevertheless, such approaches might be implemented as Netlets as well—using the Netlet Editor is just one way for Netlet design.

The goal of this tool is to ultimately link design time activities with the actual execution environment. Most of the building blocks which exist as models within

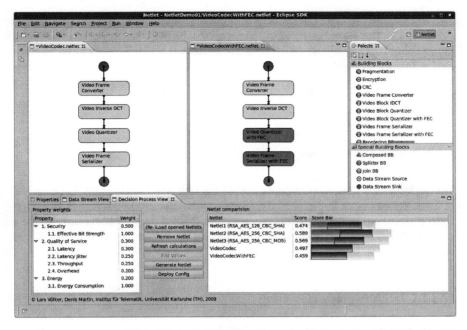

Fig. 12.1 Screenshot of Netlet Editor showing two video codec Netlets built with different building blocks

the editor have an implementation within the Node Architecture prototype described in Sect. 12.2.2. The composed Netlets designed with this tool can be created for the Node Architecture prototype in a semi-automated way. Although this is still work-in-progress, first Netlet configuration generators exist.

12.2.1.1 Implementation

We aim at a close coupling of design-time tools and the run-time environment to allow for an integrated solution for rapid protocol development. The design tool features reuse of basic building blocks for protocols that are composed in order to create new Netlets. Such building blocks provide, for instance, fragmentation of data units, or the computation of a CRC checksum. The protocol composition tool was implemented as an Eclipse plug-in based on the Graphical Editing Framework (GEF) [10] (Fig. 12.1). Its main feature is the automated aggregation of building block properties (e.g., added delay due to processing, energy consumption, etc.) as described in [35]. This allows the protocol designer to estimate how a composed protocol will act and if constraints are met early on, i.e. before implementation and simulation. It also allows him to compare the estimated performance with other solutions before he actually deploys it. Those solutions could be alternatives that are either designed by the same network architect or that come from a commercial or public third-party source.

The design tool is used to create a configuration file with instructions on how to compose a Netlet. Based on this file, the respective Netlet is automatically built from pre-existing building blocks. After that, the generated Netlet is ready to be instantiated within the Node Architecture. Due to the protocol-independent, requirements-based application interface, existing applications do not need to be adapted in order to use the new Netlet (see Sect. 4.4).

12.2.2 Running—Node Architecture Prototype

The architectural concept for future network nodes and end-systems as described in Sect. 4.4 has been implemented as a prototype. The prototype serves as a platform for running protocols of different architectures. In addition, its support for a simulation environment allows easy testing and evaluation of new protocols and architectures.

12.2.2.1 Implementation Overview

For optimal performance and code portability, C++ was chosen to implement the Node Architecture framework. This allows the user to test the code in user space first and to easily integrate it into C/C++ based simulators, such as OMNeT++ v4 [34].

The Node Architecture daemon is the core system that provides the basic components such as a Netlet repository, the Netlet Selector, and the Network Access Manager. It interfaces with the system wrapper of the respective target system and it is able to load and instantiate architecture specific multiplexers and Netlets. Both of the latter are compiled as shared libraries, which in turn are loaded by the daemon.

System Wrapper   With target system, we refer to the system environment where the daemon will be run. A system wrapper provides the necessary abstractions to access services of the underlying operating system (such as timers, threads, network send/receive, . . .). The current implementation focuses on wrappers for OMNeT++ and real systems like Linux, BSD, or others where the Boost ASIO library is available. This allows testing and evaluating the implementation within a simulator and to run the very same code on real systems.

The OMNeT++ system wrapper implementation for the Node Architecture also optionally supports the INET/INETMANET Framework [17] and the Mobility Framework 2 [24]. Thus, any topologies and mobility scenarios existing for those frameworks can be also used for the Node Architecture.

For demonstration setups, a network adapter is realized using an UDP socket, allowing it to run completely in user space. Because the Node Architecture implements the protocol handling internally, technically it does not need to make use of the TCP/IP stack of the operating system. Thus, a PCAP [27] wrapper was created as an alternative method of transmitting and receiving raw packets, being capable to avoid completely the TCP/IP stack of the operating system. Data frames are injected at the closest point before the OS passes the packets to the network interface card driver, and they are also collected right after they are received by the network driver. Although the Layer-2 MAC protocols cannot be bypassed this way (since most of the functionality is realized within the network interface card's firmware), this provides the closest possible emulation of data transmission in future networks using existing network equipment today.

Message Processing   Entities within the Node Architecture communicate with each other via special message passing interfaces. This concept was mainly introduced to support multi-threading transparently to the Netlet and building block implementations. Depending on the needs and/or properties of the host system, the number of threads can be dynamically chosen.

Netlets   Currently, several Netlets are available for this prototype: some related to data transport others to routing and signalling. For data transport, a simple transport Netlet was built that basically only adds a header for application multiplexing. For a demo showing the transport of a live video, two special video transport Netlets were built: one optimized for low bandwidth, another for more robust video transmission using forward error correction (FEC) information. Both Netlets consist of four building blocks, whereas only two are different.

Routes can be discovered automatically using one of the routing Netlets. A simple one that only discovers its direct neighborhood and two optimized ones for mobile ad-hoc networks exist: AODV (Ad-hoc On-Demand Distance Vector protocol [29]) and OLSR (Optimized Link-State Routing protocol [7]). All of them can be used interchangeably and provide forwarding information for the transport Netlets mentioned above.

Support for Mobility and Dynamic Adaptation   A strong focus of the prototype lies in mobility related scenarios. This is supported on one hand by the OMNeT++ Mobility and INET Frameworks, and on the other hand by the AODV and OLSR implementations (see sections above). As ongoing and future work, we will analyze the benefits of the Node Architecture as a programming framework when comparing and migrating between protocols: based on a given mobile ad-hoc scenario, a reactive routing protocol such as AODV may be more suitable then a proactive one such as OLSR. The necessary information in order to make an appropriate decision can be gained by simulating large networks if analytical estimations prove to be too vague. This information could, for instance, be attached to the routing Netlets and, when deploying the Node Architecture Daemon, the "best" routing Netlet could be chosen depending on the actual scenario it is facing.

Such a decision is normally only done when connecting to a new network. But since ad-hoc networks are dynamic by definition, the current network scenario may change during an active connection. This may result in the need to tune parameters of the respective protocols during run-time, e.g. to reduce the time between Hello messages in order to provoke more up-to-date information. Thus, issues for further studies include the design of tuning/optimization interfaces, algorithms for fine-tuning protocol-specific parameters, and the evaluation of those.

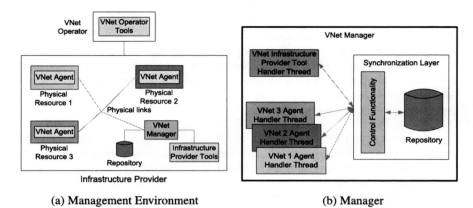

(a) Management Environment   (b) Manager

Fig. 12.2 VNet Management (VNM)

12.2.3 *VNet Management Environment Prototype*

The VNet Management (VNM) prototype is built to evaluate and demonstrate the VNet concepts and processes described in Sect. 4.9. The typical architecture of such a VNM environment consists of agents that reside in different physical resources that control the resource to create, remove or modify virtual resources. These agents are controlled by managers located centrally or distributed, which accept commands from an Infrastructure Provider to manage the physical resources. The managers hold repositories that are continuously updated based on the current status of the VNM environment. The interactions of agents and managers are similar to the operations of SNMP, but differs from SNMP due its independence from specific networking technologies (such as IP).

Figure 12.2a shows an example of a VNet management environment consisting of 3 physical resources. Each resource has a VNet Agent running in it to control the resource to manage virtual resources instantiated in the resource. The VNet Manager controls all the VNet Agents and the repository. Using a set of tools, the Infrastructure Provider initiates management requests based on the requests of the VNet Operators.

VNet Manager The VNet Manager (Fig. 12.2b), a centralized component, coordinates all the activities of the whole VNet environment. It is a multi-threaded daemon that can reside in the hardware of the Infrastructure Provider and communicate with the other components of the environment. It has the following functional activities:

- Handling of Infrastructure Provider tool connections
- Handling of VNet Agent connections
- Handling of synchronized control of requests and information arriving from or destined to the VNet Agents and Infrastructure Provider management tools

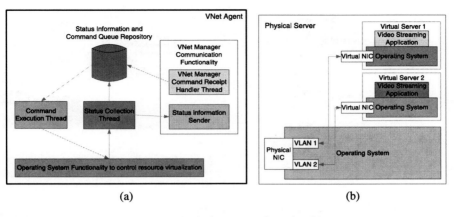

Fig. 12.3 Agent architecture (**a**) and physical server configuration (**b**)

- Handling of synchronized updates to the VNet Repository which serves as a central storage for the configurations of the VNets

All the Infrastructure Provider management tools use the same interface to send and receive data to the VNet Manager. Each tool instance started will result in a separate thread being created in the VNet Manager to handle the requests and return information. The communication between the user tools and the VNet Manager is based on TCP sockets to maintain reliable communications.

The connections of VNet Agents also have a similar concept where each VNet Agent instance will result in a thread being created to handle the sending and receiving of information to the VNet Agent. The interfaces to the agents have the same format for every agent and uses TCP/IP sockets for communications.

The control functionality which is protected (made thread safe) for synchronized access provides the following functionality to the different threads created:

- Handle command actions received from the infrastructure provider tool threads
- Command VNet Agents to perform activities
- Handle information received by the VNet Agents through the threads
- Send information to the infrastructure provider tools
- Update the repository

12.2.3.1 VNet Agents

VNet Agents manage the actual virtualization of physical resources. A VNet Agent is executed on each physical resource. Each VNet Agent is specific in its functionality to the resource under its control. Figure 12.3a shows the generic architecture of a VNet Agent.

Every VNet Agent has an interface with the VNet Manager to get commands to create, remove, modify, bring up or shut down virtual resources associated with

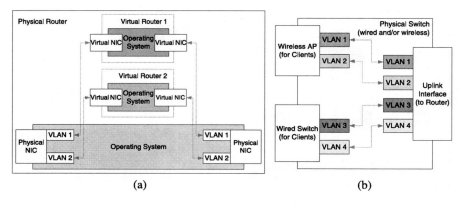

Fig. 12.4 Physical router (**a**) and switch (**b**) configuration

a given Agent. There are a number of threads in an instance of an Agent which perform the following tasks:

- Queue incoming virtual resource management instructions
- Execute these instructions in a FIFO manner
- Retrieve current status information
- Notify current status information

The operating system functionality component in the Agent architecture is the component that holds functionality that is unique to each of the resources. That means, e.g., that an Agent for a Wireless Access Point (WAP) of a particular hardware manufacturer differs from the Agent for a WAP of another hardware manufacturer in its operating system functionality component. There are a number of different VNet Agent types supported by the VNet environment.

The VNet Agent for Servers manages the virtual servers that can be created, removed, brought up, etc. on a physical server with a single network attachment. Figure 12.3b shows a VNet Agent for a physical server that has created two virtual images of servers.

The virtualization at the network level is achieved through the use of Virtual Local Area Networks (VLANs) where each VNet is carried with a separate VLAN ID.

The VNet Agent for Routers controls physical routers to manage the virtual environment. A physical router consists of multiple network interfaces to perform the routing. These interfaces are bridged to the virtual router instances to perform the routing. VNets are realized using VLANs. Figure 12.4a shows a router that has two virtual routers instantiated.

VNet Agents for Switches/Wireless Access Points use VLAN and the capability of multiple virtual SSIDs to distribute traffic to different virtual network. Figure 12.4b shows a physical Wireless Access Point handling network traffic related to four VNets.

12.2.4 Deploying—Virtual Networks

In order to verify and test the Node Architecture prototype, it has been deployed using the VNet management environment described in the previous section. This setup has the following properties: (1) The Node Architecture will run in the virtualized environment, its Network Accesses provide direct connectivity to the virtual network. (2) Multiplexing between VNets and substrate traffic is handled by the virtualization infrastructure and is transparent to the Node Architecture Daemon.

Figure 12.5a illustrates a logical view of the Node Architecture deployed in a VNet management environment using Xen based virtualization [1]. Xen is a virtual node monitor used in a number of computing hardware platforms to virtualize physical resources such as RAM, CPU time, and I/O devices. Each virtual node can obtain part of the resources of the underlying physical machine and multiple virtual nodes are running in parallel based on the scheduling mechanism of the hypervisor. Logically, one virtual node is acting just like a normal physical node. This means the Node Architecture can be deployed into any of the virtual nodes without any modifications on the daemon itself. The Network Accesses of the Node Architecture are mapped to virtual network interfaces that are bridged using VLAN.

The scenario of this small demo testbed is shown in Fig. 12.5b. The physical network basically includes one server, one router and one access point including a switch. Multiple virtual networks can run on this physical network in parallel. This means, the server and router nodes are virtualized and support more than one virtual server/router running on it. The access point supports multiple SSIDs for the virtualization of the air interface.

The main feature of this prototype testbed is to illustrate the different roles involved in a virtual networking scenario. This is achieved by graphical user interfaces representing the management terminals of the Infrastructure Provider and the VNet Operators: Requests on changing the configurations of a virtual network are issued from the VNet Operator's terminal and appear on the Infrastructure Provider's terminal to wait for approval. Once the requests are approved, the virtual network of the VNet Operator is changed. Those changes can include the addition/removal of virtual nodes and links and the change of resources assigned to the respective virtualized nodes and links.

12.2.5 Conclusion

The on-going prototype implementation of the Node Architecture Daemon including a simple example architecture resulted in a demonstration shown at the LCN 2009 [21] and was awarded the best demo based on attendee voting: Using a simple video application, we showed the impact of degrading network conditions that led to a deteriorated service quality. This required further measures to be taken either in the application or in the communication protocol in order to cope with

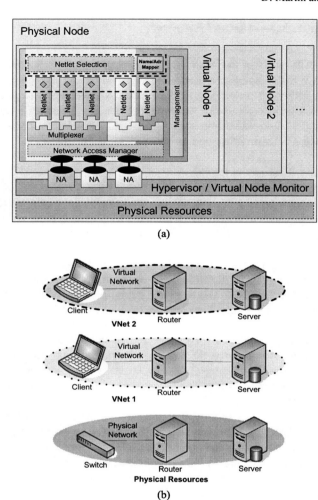

Fig. 12.5 Network virtualization setup: (**a**) Node Architecture Daemon running inside a virtual node, (**b**) Virtualization demo testbed

the changed situation. Rather than adapting the application or introducing complex, adaptive protocol features, we showed how we re-designed and enhanced the simple Netlet used before with further protocol mechanisms using the Netlet Editor. The enhanced Netlet was then deployed on the Node Architecture and the selection process decided to use the new Netlet since it provided a more robust data delivery.

The demonstration setup was continuously enhanced: Using a small virtualization testbed, the video streaming scenario was seamlessly transferred into a virtual network. The virtualization testbed, shown at several project-wide events, illustrated the roles of the different stakeholders and was able to show the feasibility of resource reservation and isolation of virtual networks.

The key features implemented by this prototype setup include:

- Rapid creation of composed protocols (Netlets) using a graphical design tool
- Reuse of existing protocol building blocks
- Automatically selecting an appropriate Netlet based on application requirements and network properties experienced at end nodes
- Isolation of virtual networks
- Separation of roles in virtual networking scenarios

To broaden the basis and get more diverse results, our future plan is to include attachment of end systems to multiple virtual networks, and to test the daemon also in mobile ad-hoc scenarios. Some aspects will be further developed within G-Lab [12], a German testbed platform distributed among multiple locations. The code of the Node Architecture prototype is released under the conditions of the GNU General Public License (GPLv2) and is available online [22].

12.3 Network Virtualization Architecture Prototype

The prototype presented in this section implements several components of the Network Virtualization Architecture described in Sect. 4.9, allowing the provisioning of customized virtual networks (VNets) on top of multiple Physical Infrastructure Providers (InPs) on-the-fly. In contrast to the VNet Management prototype presented in Sect. 12.2.3, this implementation focuses on a larger scale prototype within a testbed environment. It essentially implements and fully automates the VNet lifecycle as discussed in this book and in [32]. We have been providing the means for the feasibility study of the Network Virtualization architecture, showing the performance and its scalability within a medium-size experimental infrastructure (see [28, 32]).

12.3.1 Infrastructure and Software

The prototype is implemented on *Heterogeneous Experimental Network* (HEN) [16], which includes more than 110 computers connected together by a single non-blocking, constant-latency Gigabit Ethernet switch. We mainly use Dell PowerEdge 2950 systems with two Intel quad-core CPUs, 8 GB of DDR2 667 MHz memory and 8 or 12 Gigabit ports.

We take advantage of existing node and link virtualization technologies for the provisioning and management access to the VNets. We use Xen 3.2.1 [1], Linux 2.6.19.2 and the Click Modular Router [20] (version 1.6 but with patches eliminating SMP-based locking issues) with a polling driver for packet forwarding. We rely on Xen's paravirtualization for hosting virtual nodes, since it provides adequate levels of isolation and high performance, as shown in [11].

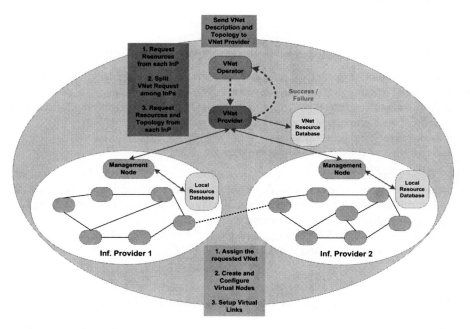

Fig. 12.6 Prototype overview

12.3.2 Prototype Overview

Figure 12.6 illustrates an overview of the prototype. A fixed number of HEN nodes compose the substrate which is split into multiple logical clusters, each one serving as an independent InP. InP topologies are constructed off-line by configuring Virtual Local Area Networks (VLANs) in the HEN switch. This process is automated via a *switch-daemon* which receives VLAN requests and configures the switch accordingly.

Separate physical nodes act as the VNet Provider (VNP) and VNet Operators (VNO). VNP establishes direct connection with dedicated management nodes belonging to the InPs. These management entities handle all the VNet requests on behalf of their corresponding InP. VNO, VNP and all substrate nodes expose provisioning and console interfaces which allow remote procedure calls based on XML-RPC.

VNet requirements are formulated and relayed from the VNO to the VNP. Such requirements describe the number of virtual resource instances, as well as their respective properties. Since VNet requests are not known in advance, they are processed as they arrive both by the VNP and the InPs. The outcome of each request is communicated to the VNP and subsequently to the VNO. All VNet requests are communicated using an XML-based resource description model which includes separate descriptors for nodes and links, allowing ultimately a coherent VNet specification.

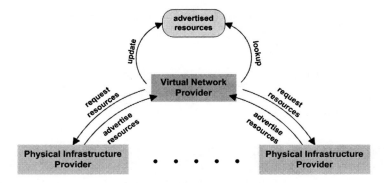

Fig. 12.7 Resource discovery with limited information

The prototype implements all the required VNet provisioning steps including: (i) resource discovery, (ii) VNet assignment, (iii) virtual node instantiation and configuration, and (iv) virtual link setup. To handle multiple InPs and the possibility of limited information disclosure, we provide resource discovery systems with different levels of information disclosure and an algorithm for splitting incoming VNet request among the InPs based on contractual agreements or requests that should be fulfilled (e.g., location, CPU, bandwidth, number of physical interfaces, etc.). Further details on VNet provisioning are given in Sect. 12.3.3. Upon VNet instantiation, the prototype establishes management access to the virtualized nodes, allowing VNO to operate the VNet (see Sect. 12.3.4).

12.3.3 Virtual Network Provisioning

VNet instantiation involves several interactions between the VNO and the VNP, as well as the VNP and the participating InPs. Essentially, the instantiation of a VNet takes place in the following consecutive steps:

1. Resource discovery: A critical parameter affecting resource discovery is the level of information disclosure allowed by the InPs. Our implementation provides two systems for resource discovery when limited or no information is exposed by InPs, respectively. Figure 12.7 illustrates how the VNP interacts with the InPs to collect the resource information advertised by them. The VNP updates and maintains all advertised resource information, on which it relies for splitting VNet requests among InPs.
   It is possible that some InPs might be unwilling to expose any resource information. Since the resource discovery system (RDS) in Fig. 12.7 discounts this possibility, we provide an additional RDS which is based on resource queries invoked by the VNP. InP responses to such queries may vary depending on the level of information disclosure.

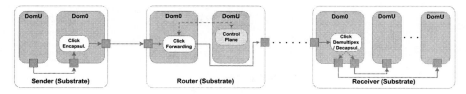

Fig. 12.8 Virtual link setup

2. VNet embedding: Embedding each partial VNet to the respective InP is essentially decomposed to node and link assignment. The primary goal for our node assignment strategy is to maintain low and balanced stress (i.e., number of hosted virtual nodes) among the substrate nodes. Following node assignment, each requested virtual link is mapped to a substrate path based on the shortest-path algorithm. To this end, a cost per substrate link (which represents the link stress) is maintained and used by the algorithm for path selection.
3. Node setup: Upon VNet assignment, each management node signals individual requests to the substrate nodes. Each substrate node handles the incoming request within its management domain (Dom0), triggering the appropriate action (e.g., virtual node creation/configuration).
4. Virtual link setup: For the inter-connection of the virtual nodes, we currently use tunnels with IP-in-IP encapsulation, as shown in Fig. 12.8 (other tunnelling technologies can be used as well). Each virtual node uses its virtual interface to transmit packets, which are captured by Click Modular Router for encapsulation, before being injected to the tunnel. On the receiving host, Click demultiplexes the incoming packets delivering them to the appropriate virtual node. For packet forwarding, we use Click SMP with a polling driver. In all cases, Click runs in kernel space. Substrate nodes that route packets consolidate all virtual forwarding paths in a single domain (Dom0) preventing the costly hypervisor domain switches; hence, packet forwarding rates are very high (see [11, 28]).

This sequence of steps provides a fully virtualized network which is ready to be operated and managed by the VNO.

12.3.4 Management Access

To enable VNO to operate any instantiated VNet, we provide management access to all virtual nodes. Management access is relayed via the VNP to the proper InP and subsequently to the physical node that hosts the target virtual node, as shown in Fig. 12.9. To allow separation between multiple VNets, we use a globally unique identifier for VNets, namely *vnetID*. We also use the identifier *vmID* for the virtual nodes. The scope of *vmID* is restricted to a specific VNet. Note that *vmID* provides separation among virtual nodes from the view of VNO and in principle, it does not match the identifier (i.e., *vmconfigID*) used by a particular node virtualization

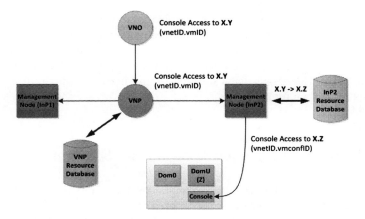

Fig. 12.9 Management access to virtual nodes

technology (e.g., Xen) within an InP. Translation between *vmID* and *vmconfigID* is provided by the InP management nodes which invoke lookups to the InP resource databases.

VNP, InP management nodes and all substrate nodes run daemons which expose dedicated interfaces for management access. As depicted in Fig. 12.9, a VNO request for management access to the node Y of VNet X is first sent to the VNP which has to determine to which InP the request needs to be relayed. To this end, VNP performs a lookup to its resource database and matches $X.Y$ with one of the participating InPs. Subsequently, the InP management node translates vmID Y to a vmconfigID Z and uses $X.Z$ to locate the physical node that hosts the target virtual node. Following these steps, the VNO is in position to establish management access to any virtual node and apply the desired configuration to a VNet.

12.3.5 Conclusions

We implemented a prototype that enables the provisioning and management of customized VNets on top of multiple InPs, in accordance with the network virtualization architecture described in this book (see Sect. 4.9). This prototype also provides insights into critical architectural design decisions, such as the tussle of information disclosure against information hiding.

Despite the increased complexity for resource discovery and allocation, we used this prototype to show that VNets can be provisioned within a few seconds [28, 32]. Our experimental study did not uncover any scalability issues within our infrastructure, even in the presence of multiple InPs. Our results indicate that new business models can create commercial products that setup VNets in large infrastructures in the order of minutes, while complying with restrictions imposed by InPs such as limited resource knowledge.

With our prototype implementation, we also showed what technological ingredients are needed and how these can be combined efficiently to provision and manage VNets. The main building blocks of our prototype, such as node virtualization technologies and packet encapsulation/decapsulation for virtual link setup, are readily available today. This shows that a shift toward full network virtualization is viable in the not too distant future.

12.4 Real-Time Adaptation in Emergency Scenarios

This prototype applies the new architectural concepts of this book for the deployment of self-managed, large scale networks. The prototype proves the validity of the software engineering techniques proposed in Sect. 4.2, the interoperability of families of network architectures, and the enhanced performance of an in-network management design described in Chap. 8.

12.4.1 Prototype Elements

The prototype is integrating different elements from several architectures described in this book, namely the strata concept, the Component Based Architecture (CBA), interoperability functions, the in-network management framework, and in-network management algorithms.

12.4.1.1 Strata

With strata taking a macroscopic view of the Architectural Framework, it goes beyond the concept of OSI layering and supports the design of network architectures following a modular and flexible "black box" approach; that means that specific network functions can be implemented over any software or hardware platform thanks to the specification of the proper external interfaces.

The strata architecture is composed of a set of generalized strata, each playing a specific role in and across networks. For example, Knowledge and Governance strata (defined in Sect. 4.2) are instantiated at this level.

The Bootstrap stratum component plays a special role during the startup of the nodes in the network, as it aids in the initiation of the operation of firstly the Machine stratum, and then secondly of the Knowledge and Governance strata. From that point on, strata as well as the domain as a whole are self-managed. In this way, the strata architecture provides for this prototype the overall framework and interfaces for self-managed network functions, as well as the auto-creation of network domains and their composition.

12.4.1.2 Component Based Architecture (CBA)

In order to support the encapsulation of constituent network features as the unit of independent deployment in a stratum, a network component needs to be concretely separated from its environment and other components. The Component Based Architecture (CBA) plays the role of defining these components, their relationships and their functionalities within the design repository of the architectural framework. At the higher abstraction levels of Strata and Netlets, CBA is providing the functional blocks (components) and units of interoperability (contracts) that can be used during the composition process.

The CBA prototype has developed and deployed a completely modular componentized and functioning IPv4 network stack. The stack components (ARP, RARP, IP, Naming, TCP, UDP) are developed in Java and originate from the JNode OS [18]. With the addition of the SocketAPI, NetworkDeviceManager, and NetworkLayerManager, this prototype now offers each component functionality as a services within an OSGi [26] environment. This allows components to implement a service orientated network stack.

This is not a typical network stack configuration as within almost all operating systems today, where the more traditional approach is a monolithic configuration. The CBA network stack provides three main features:

(a) It virtualizes the network interface hardware and the operating system support.
(b) It provides transparent integration of the stack to existing applications via the socket API.
(c) It provides a simple development and deployment environment.

CBA also provides a standard design, develop and deployment process which allows the designer to select the components required, assign existing components or develop new ones, assign these components to a network node, and then deploy these node(s) as required to a virtualized network.

12.4.1.3 In-Network Management

The In-Network Management (INM) described in Chap. 8 is an enabling approach for decentralization, self-organization, and autonomy which are the only mechanisms which will support the effective management of large networks as foreseen for the future Internet. The basic idea is that management tasks are delegated from management stations outside the network to the network itself and therefore intelligence is embedded in the network. This allows a network node to execute management functions on its own, allowing for self-reconfiguration or self-healing in an autonomic manner.

In order to realize this vision in a prototype, an INM framework has been created with processing and communication functions associated with each network element or device. Those elements or devices communicate with peer entities in its proximity. In addition, they are able to monitor and configure local parameters. The collection of these entities creates a management plane, a thin layer of management functionality inside the network that performs monitoring in the Knowledge stratum and control tasks in the Governance stratum. INM entities are the elements of the Governance and Knowledge strata. Which INM entity to deploy is therefore part of the design phase of those strata.

In this prototype implementation, the INM framework provides the minimal functions common among INM algorithms to support self-organization: a naming scheme for INM algorithms, discovery, organization interface for control of objectives and a visual interface for operators. From a functional point of view, the prototype integrates real-time monitoring of network-wide metrics and local triggering of self-optimization: these functions are applied to the case of management of dynamic traffic fluctuations in the network and healing of congestion.

Thanks to the gossiping protocol "INM Monitoring" a network wide congestion status can be estimated in a cheap manner by means of preserving resource consumption. In case of high network congestion meaning that a pre-configured congestion threshold is crossed, "INM Monitoring" autonomously invokes the INM algorithm "Congestion Control" based on emergent behavior. By then, network congestion is efficiently distributed and balanced via multiple possible routes and if needed the congestion status of each router and available path visualized in a detailed view.

INM implements distributed management functions with reliable behavior. To achieve this important feature, key properties of distributed systems are part of the design (e.g. performance, timeliness, etc.) and then part of the management interfaces (e.g. through objective over the organization interface). The control of these properties is an important requirement to the transfer of INM to operative networks in the future Internet and has been therefore subject of the proof-of-concept through the prototype. Figure 12.10 shows how an operator would then be able to control these properties through high-level objectives (e.g. level of autonomicity or timeliness of self-optimization) and how this information can then be mapped to the selected functions in the prototype.

In order to support the concept of interoperability the prototype also implements a Service Level Agreement (SLA) manager which provides a contract mechanism with meta-data and semantics related to the network component interfaces and functionalities. This component contract forms the basis for interoperability and composition, and allows for SLA to be negotiated across heterogeneous networks along with validating that the SLA is not violated.

In the prototype, the SLA Manager module controls the interdomain SLA management process. It establishes a communication channel to SLA Managers in discovered neighboring domains. The SLA Manager is responsible for negotiating SLAs with neighboring domains. It also takes care of processing measurements received from the Knowledge stratum and eventually notifying the Governance stratum when preprogrammed alarm thresholds are violated; as a concrete example, a congestion control module can be notified. It automatically renegotiates SLAs with neighbors according to a preprogrammed policy when preprogrammed thresholds for unacceptable levels are violated.

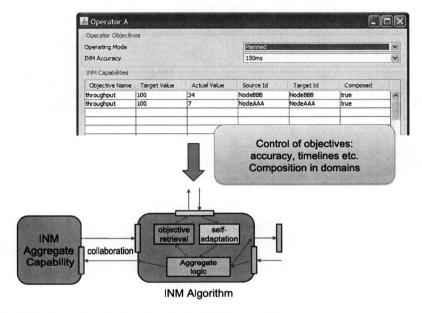

Fig. 12.10 Control of objectives through the INM Framework

12.4.2 Implemented Scenario

From an application point of view, the prototype enables real-time management of large-scale networks. A major scenario where the selected functions can be deployed in the future Internet is an emergency scenario, which is characterized by dynamicity and the need for timely adaptation. For example, the network can be thought in conditions after a natural disaster, where emergency teams need to provide relief, but the network infrastructure is heavily damaged. Multiple network domains can be bootstrapped, and a SLA negotiation is undertaken to transit the medical team node traffic between two domains.

The prototype bootstraps four distinct domains: (i) Fixed Operator (FO) Network, (ii) Emergency Medical Team (EMT) Network, (iii) Ad Hoc Disaster Recovery (AHDR) Network and (iv) Emergency Response Command (ERC) Network. This scenario is depicted in Fig. 12.11.

The challenges in this scenario are related to the establishment and maintenance of a good Quality of Service to guarantee a reliable connection to the emergency team. Traditionally, the Internet would be used as instrument for reliable connection in emergency conditions, but legacy phone lines would be reserved. The major limitation is that instruments for dynamic reconfiguration are not put in place in the network. The enhanced architecture presented here increases the performance of such a network. The Knowledge and Governance stratum are used to enforce the necessary control loops on various areas of the network. Monitoring algorithms implemented in the INM entities assure the necessary scalability to control in real-

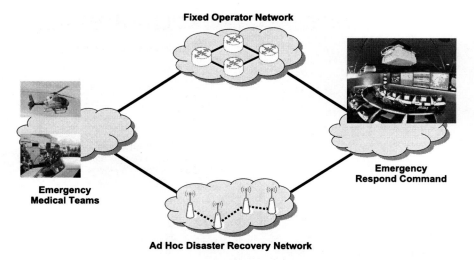

Fig. 12.11 Reference scenarios for application of the prototype in the future Internet

time a large scale network like the one depicted in the scenario. The SLA manager is in charge of the negotiation between different domains. The integration of the distributed management capabilities within the CBA architecture guarantees reliability in the emergency.

In order to achieve this, the "Strata" enables dynamic domain membership, while the "INM Framework" enables discovering management functionality in a domain and the "SLA Manager" enables negotiating services with specific QoS parameters between domains.

For the SLA negotiation, the "SLA Manager" performs inter-domain negotiation (showing interoperability) thus creating a SLA between the EMT network and the FO network on behalf of the client node of one of the medical teams. The "INM Monitoring" module of the prototype is used to monitor SLA compliance.

The network is now in a steady state, however to highlight the dynamicity and the need for timely adaptation in the network, simulated congestion build up is applied to the FO network. The "INM Monitoring" detects this congestion increase in the FO network, which kick-starts the "Congestion Control" module of the prototype, which works to mitigate this congestion through routing strategies. At the same time the "SLA Manager" in the EMT domain, prepares a contingency plan by negotiating a back-up service with the AHDR network.

With a dramatic increase in congestion built up in the FO network the "INM Monitoring" module notifies the Knowledge stratum of escalated congestion. With this the Governance stratum acts on the Knowledge stratum and informs the "SLA Manager" of the congestion, and as there is a SLA violation the "SLA Manager" in the EMT network decides to switch the service level agreement from the FO domain to the AHDR domain.

12.4.3 Conclusion

Having completed his task, the real-time adaptation in the emergency prototype has shown the self-organizing and self-deployment of a network domain through the holistic and systematic approach of the strata architecture. The prototype also shows how CBA provides a service orientated architecture based network stack, which allows for more flexible and modular networking functionality. This approach provides the following advantages to network developers (protocol or functions):

(a) Java environment which is not tied to any operating system,
(b) shortened revise/test cycle (no kernel reboots),
(c) easier debugging—full access to the latest java application toolsets,
(d) improved stability—user level development, unstable protocol components only affect the application not the overall system.

The INM framework and algorithms for monitoring and congestion control provide distributed domain level congestion detections with optional self-healing capabilities while also allowing domain owners to manage their domains at an objective level.

Finally interoperability via SLA negotiation at a domain level is highlighted in the prototype with the SLA switched networking enabled via Strata encapsulation of the differing network architectures at a domain level.

12.5 Integrating Generic Paths and NetInf

This prototype consists of two separate components that serve as proof of concept implementations for the Network of Information (NetInf) architecture (Chap. 10) and the Generic Path (GP) architecture (Chap. 9). Both components can be used on their own to implement and evaluate aspects in their respective areas or can be combined to a powerful prototype testbed that covers information-centric networking and advanced data transport mechanisms at the same time. This way, benefits of the tight integration can be demonstrated, like NetInf-adapted data transfer mechanisms that require functionality in intermediate nodes, e.g., for caching of data and meta-data.

12.5.1 Generic Path Prototype Implementation

This section introduces the prototype implementation of the *Generic Path (GP) architecture* (Chap. 9).

In the GP architecture, we chose an object-oriented approach to design network components while keeping them coherent in their interfaces and basic structures. This allows to incorporate new networking techniques more flexibly than in today's network architectures as networks can be arbitrarily composed of the components. Furthermore, the composition of networking functions can be easily adapted according to any cross-layer information during runtime, thanks to the unified interfaces. Examples for data transport aspects that have been modified/integrated into this architecture are routing, mobility [2], cooperation and coding techniques [3], and resource allocation.

The basic building blocks that constitute the GP architecture are Entities, Endpoints, Hooks, and the Core. Their role in the overall architecture has been introduced in detail in Sect. 9.2. The following sections discuss the main design choices that have been made during the GP prototype development.

To be able to use the prototype, i.e., Entity, Endpoint, and Core implementations, in various environments, like Linux, Windows, and embedded systems, we used C++ for efficiency and strictly separated the implementation of the logic parts from the environment-specific parts. In detail, this separation means that our testbed abstracts execution environment functions, e.g., by mapping Hooks to available IPC mechanisms. Specialized Entity and Core implementations, like a Peer-to-Peer Entity, just use these abstractions. This way, the Entity and Core implementations, i.e. the contained protocols, routing strategies, mobility schemes, data encoding, etc., can be used in different environments without the need for adapting them, i.e., without touching their code.

We realized this separation by inheritance, provided by the object-oriented programming paradigm. For this, we implemented all abstractions of the execution environment like timeout and callback handling in a root class, called `Abstract-TimeoutManager`, and all GP architecture features like Hook handling, Endpoint creation, and name resolution in `AbstractCore` and `AbstractEntity`. From `AbstractEntity`, wrapper classes like `PosixEntity` and `OmnetEntity` inherit to map the abstractions to the appropriate execution environment APIs. Thereafter, an `Entity` class is derived from *one* of these wrappers (chosen during compile time). New, environment-independent Entity classes inherit from this class. Environment-specific entities directly inherit from a wrapper class. Figure 12.12 illustrates the inheritance relations.

Currently, we have ready-to-use wrappers for POSIX-compliant systems, like Linux, BSD, and Darwin (Mac OS X), for Windows, and for OMNeT++ [34], an open-source discrete event simulator. The wrapper for OMNeT++ is especially useful during the development phase and for demonstrating scalability while at the same time using the implementation in real-world scenarios via the POSIX/Windows wrapper. Future work includes an implementation for Open-Flow [23].

For Endpoints, a similar inheritance graph exists. The root class only contains the basic GP functions that are common to all possible Endpoint types. This means that these functions are independent of any communication paradigm (e.g., publish/subscribe vs. send/receive) and independent of any Compartment in which a derived Endpoint is instantiated eventually. Hence, the Endpoint root class mainly contains management functions that are common to all Endpoint types.

Besides the basic GP architecture elements that focus on data transport, the prototype implementation also fully implements the name resolution framework that

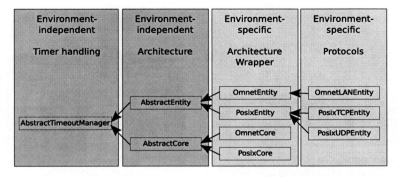

Fig. 12.12 Inheritance graph for Entity and Core implementations. In this example, all Entities implementing protocols (leaves of the tree) are environment-specific as they require some functions of a certain execution environment (OMNeT++ LAN interface and POSIX TCP/UDP stack). Environment-independent Entities would inherit directly from `AbstractEntity`

has been introduced in Chap. 5. This way, the prototype provides all features to construct complex networking systems. More information on this prototype can be found in [5, 6].

12.5.2 NetInf Prototype Implementation

The NetInf prototype focuses on two main aspects: on the development of so called *NetInf nodes* (*NINs*) and on the architecture of an overall Information-Centric Network (ICN) that is composed of NetInf nodes (Fig. 12.13a). In general, each NetInf node can perform the role of a server as well as a client for information-centric requests. Each NetInf node provides an information-centric Application Programmer Interface (API), called *NetInf API*. The main primitives include searching for information and resolving information identifiers (IDs) into appropriate content locators. Thereby, NetInf nodes can build, e.g., an ad hoc network and can perform information-centric requests on neighboring NetInf nodes.

In addition, our prototype integrates specialized NetInf nodes that can provide services like search and lookup on a global scale, thereby providing an infrastructure that can be used by other nodes (clients) to easily implement various information-centric applications.

The NetInf prototype is consistently based on the interface design pattern for flexibility. It is implemented in Java to allow for platform independence and a flexible choice of devices. Special attention has been paid to the selection of libraries to make the framework usable also on mobile devices. The code currently works on FreeBSD, Linux, Windows, and Android.

The plugin concept is based on Google's lightweight dependency injection framework *Guice* [13]. Guice is used for two specific purposes. First, we use Guice for constructing an information-centric node from multiple components, making the

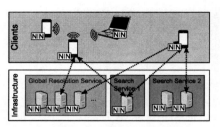

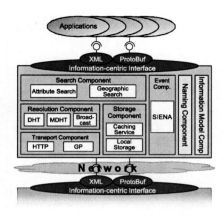

(a) Architecture overview. All participating nodes are NetInf nodes (NINs).

(b) Example NetInf node configuration, consisting of adaptable components that each contain several service incarnations.

Fig. 12.13 A Network of Information consisting of NetInf nodes

overall node architecture easily extensible and specific component implementations exchangeable. Second, Guice is used to plug component services into components to extend them with new service implementations.

In the following subsections, we will first describe the architecture of a NetInf node and will then explain how those nodes are combined into an ICN architecture.

12.5.2.1 NetInf Nodes

Figure 12.13b gives an example of a NetInf node structure. The node is composed of several components. Each component can contain a single (see *Naming, Information Model* component) or multiple different implementations of its component functionality (e.g., *Resolution* component).

Applications access an information-centric node via its adaptable interface. The same interface can be used to communicate with other information-centric nodes over the network, as illustrated at the bottom of the figure. Inter-node communication and applications will typically access the interface in different ways.

To provide a broad and flexible mechanism to access the interface, we have implemented a layer of indirection on top of the interface that provides access in different ways and can easily be extended with new mechanisms. Our NetInf prototype extends the information-centric node interface with three different mechanisms to access the node interface (Fig. 12.13b): Java applications can simply use the provided Java interface. For applications that talk HTTP (e.g., a Web browser plugin), we provide an HTTP proxy interface. For inter-node communication, we provide access to the node interface via Google Protocol Buffers (Protobuf) [14]. Google Protobuf enables the simple and fast definition of custom protocol messages and provides efficient data transfer. Many different languages like Java, Python, and

C++ are supported, thereby also enabling communication between network nodes written in different languages.

A NetInf node consists of the following main components (cf. Chap. 10): *Search, Naming, Name Resolution, Data Transport, Storage, Event Service*, and *Information Model*. Each component can be adapted with node-specific services and protocols based on a flexible plugin concept. Services and protocols are encapsulated in *component services*. To enhance flexibility, a component can contain multiple component services while each service fulfills the same interface but may implement and fulfill this service in a different way. A *component controller* is responsible for choosing between component services, and managing the order of execution. For example, the *Resolution Controller* manages several *Resolution Services* that implement specific name resolution mechanisms, e.g., via a Distributed Hash Table (DHT) system or via broadcast, that can be chosen depending on the context. Adding a new service to a component is done simply by plugging the new component service into the specific component and reconfiguring the controller to include the new service.

12.5.2.2 NetInf Infrastructure

In the NetInf prototype, an ICN consists of any number of NetInf nodes. Those NetInf nodes communicate with each other using the NetInf API and, thereby, form the ICN.

NetInf nodes can perform different roles in the ICN. First, there are 'common' NetInf nodes that typically perform a client role but can also provide NetInf services to other NetInf nodes. Second, there are specialized NetInf nodes that build the infrastructure of the ICN.

The prototype infrastructure consists of the *Information Lookup Service* (*ILS*) and the *Information Object Lookup Service* (*IOLS*). The IOLS is a global resolution service that performs lookup operations for clients, i.e., it returns a corresponding Information Object (IO) for a given NetInf ID. This service is the central component for a global NetInf and, hence, must scale and must have a high availability. Therefore, we implemented the IOLS based on a DHT using FreePastry [33]. Note that this Peer-to-Peer (P2P) implementation lacks important features, like low query latency, which is acceptable for our prototype. A productive implementation would use more sophisticated techniques, like the Multiple Distributed Hash Tables (MDHT) architecture [8].

The second component of the infrastructure are search services. These services perform search operations for clients, i.e., return a set of NetInf IDs for a given attribute query, like `rfid == 'A1B2C3'`. This way, IOs can be found based on their descriptive attributes without knowing their ID.

More information about this prototype can be found in [5, 6, 9].

12.5.3 Combining GP/NetInf Prototypes

To show the power of combining the GP architecture with the information-centric networking provided by NetInf, we implemented a NetInf Entity for the GP architecture. This Entity is part of the NetInf Node on the NetInf side and joins a NetInf Compartment on the GP side. This way, the NetInf node can use all features of the GP architecture, e.g., for data transfer or name resolution.

Two scenarios where this interaction is especially useful are NetInf session mobility for data transfer/streaming and bootstrapping. During its bootstrapping phase, NetInf uses the GP name resolution mechanisms to easily find addresses of its NetInf-enabled neighbor nodes. This is an essential feature for information-centric networking. It increases robustness and flexibility as it permits to exploit local NetInf neighbors in case of limited Internet connectivity.

NetInf knows about all locators from which a certain piece of data can be downloaded/streamed. Hence, when new locators become available for an ongoing transfer that are cheaper or provide better performance, NetInf can easily initiate a GP Endpoint migration. This way, the source of the data transfer changes without interrupting the transfer; everything happens transparently to the downloading node.

12.5.4 Conclusion

Both the NetInf and the GP prototype prove the benefits of the architectural concepts that have been introduced in Chaps. 9 and 10. In particular, the NetInf prototype demonstrates the specific NetInf advantages like improved resource usage, inherent load balancing, efficient data dissemination, and improved security features that are deeply integrated into the network architecture. In addition, the GP prototype shows that disposing the strict ISO/OSI layering provides a lot of flexibility. It permits to easily implement networking features that are difficult to achieve today, e.g. session mobility and dynamic use of cooperation and coding techniques, which require a lot of cross-layer information.

Integrating both prototypes has significantly improved the understanding of how both concepts interact and how they can benefit from each other. The two examples described in this section, bootstrapping and session mobility, illustrate the potential of GP's generic name resolution scheme and of building custom-tailored GPs that provide NetInf-specific transport. Future examples may include GPs that support transparent in-network caching.

We are planning to release the NetInf and GP prototype code base as open source project to make it available to a wider audience.

12.6 INM Cross-layer QoS Used in Generic Path

The prototype presented in this section combines two concepts described in this book (Chaps. 8 and 9): In-Network Management (INM) and Generic Path (GP).

The results obtained through practical implementation proved that the application of cross-layer QoS in congestion control is feasible.

12.6.1 Introduction

One major challenge in the Future Internet is to ensure a specific level of QoS for the services provided to the user. One possible solution to achieve this goal is to apply In-Network Management [25] capabilities to Generic Path [4, 15], mainly to face congestion control issues. Whenever the available transfer rate is less than a minimum threshold imposed by legacy solutions for a given link, a Network Coding (NC)-based GP could be instantiated to maintain the SLA. The implemented prototype demonstrates that this might be a solution for providers when over-provisioning and/or re-routing are not possible.

12.6.2 Cross-layer QoS Based Testbed Used for Generic Path

A new paradigm was proposed in this demonstrator in order to face the congestion. Thus each strategic node, i.e. a node that has the major management functionalities such as neighbor discovery, registry, resource control, event handling, security etc., should run a software bundle acting as a quality of service dedicated management capability, called Cross-Layer QoS (CLQ). This works with two types of interfaces: organization and collaboration, and it implements cross-layering for QoS parameters and/or data. The measurements performed between nodes are within the defined scope (intra-domain, inter-domain), accessing the hardware directly, through the collaboration interface. The CLQ was implemented using two approaches [25, 31]:

- **Bottom–up approach:** enables traffic parameters collecting like: available transfer rate, one-way delay, bit error rate. Note that they are characterizing a specific physical link and are an objective way of evaluating the hop-to-hop communication channel. The results could be obtained directly from the hardware driver where the technology permits, or using different dedicated tools performing passive or active measurements between nodes. The traffic parameters collected are exchanged between the QoS modules through the collaboration interface.
- **Top–down approach:** imposes a specific transfer rate to the hardware through INM platform and hardware driver. The traffic parameters are received from different GPs through the organization interface and are sent directly to the hardware, using the collaboration interface.

The initial version of the testbed was simulated in OMNeT++, as it was presented at IEEE SOFTCOM 2009 [30], but herein we are presenting the real-time implementation. Thus the CLQ has two parts: the measurement part, written in C, and the publishing one, running as a bundle on top of Java-based Open Services Gateway

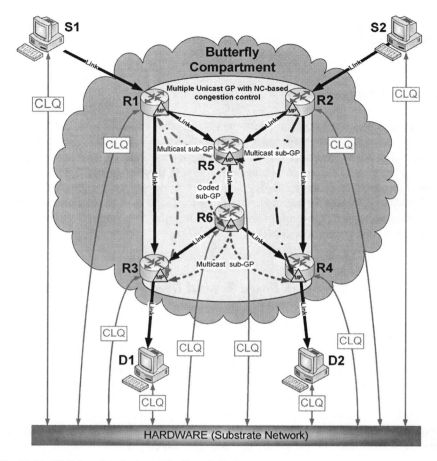

Fig. 12.14 CLQ-based testbed used for Generic Path

initiative (OSGi). The latter is needed to make the results available to other self-managing entities that are communicating through organization or collaboration interfaces. The NC-based GP module is written in C, being integrated with CLQ in all strategic nodes. For demonstration purposes only, the UDP transmission is used to interconnect the coding nodes. However both CLQ and NC could be implemented in the near future on top of MAC Sub-Layer. The testbed presented in Fig. 12.14 includes six routers with CLQ and NC capabilities (R1-R6), each one running in a Linux-based machine. The data flow generators (S1, S2) and the destinations (D1, D2) are actually represented by PCs.

Specialized software written in C under Fedora Core is running on each node to monitor the substrate resources, i.e. the transfer rates and the one-way delays between the neighboring nodes, in order to assist congestion control mechanisms, to get a global perspective, and to have statistics on link status. The NC-based GP scheme was applied in order to obtain a better congestion control, following three principles referring to instantiation, dynamic application and dynamic flow encoding. Note that the internal architecture includes sub-GPs connected by Mediation Points (MPs), as it is presented in Fig. 12.14.

The first principle refers to the instantiation of NC-based GPs [4] with multiple functioning modes. It depends on long term statistics provided by INM and used to identify network topologies where NC solutions exist and the congestion level is high enough. GPs with NC capabilities are instantiated on the identified network topologies but NC operations are initiated only when QoS-aware routing is not possible. This principle was simplified in our testbed, the network topology being fixed.

The second principle covers the dynamic NC application, meaning that the link and flow characteristics are continuously monitored by INM, triggering the situations when the short/medium term statistics of congestion level and links transfer characteristics would make NC-based GP a viable solution.

The last principle discusses the dynamic flow encoding process. Thus, theoretical studies of NC operations [19] consider that coding operations are performed continuously. Due to the burstiness of the real flows, data packets to be encoded are not present all the time in each node and coding can not be continuous. The proposed solution was the following: once the NC operations are activated coding operations are performed only in the moments when data packets are available for encoding, otherwise non-coded packets are transmitted, as the decoding algorithm is able to discriminate them. Activation of the NC operations was realized only if the topology link parameters fulfilled imposed transfer rate and delay conditions and if congestions were detected for longer periods of time. Dynamic XOR based flow encoding was performed in node R5, while the decoding was performed in nodes R3 and R4.

12.6.3 Achievements

The first achievement is the proof of real-time implementation feasibility of cross-layer QoS for new congestion control schemes based on network coding. The solution was requested by the practical situation whenever the infrastructure provider cannot perform over-provisioning or re-routing to face congestion (for instance in rural wireless networks). The second achievement is that the prototype allows to evaluate the performances comparing to more complex techniques such as QoS-aware routing.

12.6.4 Experimental Results

Two video streams R1-R5-R6-R4 (see Fig. 12.15a) and R2-R5-R6-R3 were sent by VLC clients connected to R1 and R2 and sharing the link R5-R6 that could be congested during the experiments. The performance evaluation was based on the

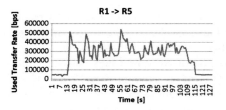

(a) Original flow sent by source node R1 to encoding node R5

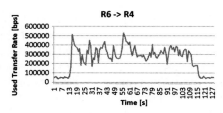

(b) Case 1: the flow received by node R4 on the path R1-R5-R6-R4

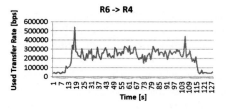

(c) Case 2: the flow received by R4 could not be used; the packet lost rate was about 20%

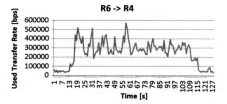

(d) Case 3: the flow received by R4 generated a good QoE; the packet lost rate was about 2%

Fig. 12.15 Used transfer rate for the flow R1-R5-R6-R4

following parameters measured separately per each video flow: number of packets lost, time distribution of packets lost, variation of the time between two consecutive video packets. Three cases were actually envisaged:

(a) Case 1 (no congestion, no NC-awareness): due to enough available transfer rate on link R5-R6, the quality of experience at the destinations was very good. This was demonstrated for the flow R1-R5-R6-R4, as in Fig. 12.15b, but similar results were obtained for the other stream too.

(b) Case 2 (congestion on link R5-R6, no NC-awareness): the shared link R5-R6 was congested (due to the background traffic injected), the available transfer rate being less than the value (about 1 Mbps) required to transmit the video streams simultaneously. Both receiving nodes R3 and R4 experienced a bad quality of the movies because of the packets lost in the congested link. The degradation of the flow received at R4 can be observed in Fig. 12.15c.

(c) Case 3 (congestion on link R5-R6, with NC-awareness): the link between R5 and R6 remained congested as in Case 2, but a NC-based GP was instantiated whenever CLQ triggered the situation. Pre-congestion was detected using the INM algorithm, the mechanism being activated in advance, before severe congestion might occur. In Fig. 12.15d the shape of the flow received at R4 is close to the original one.

The measurement results proved the efficiency of the NC-based mechanism, as the percentage of packets lost decreased from about 20% in Case 2 (making the video streaming service unacceptable) to about 2% in Case 3, with NC-awareness (quality of experience being good despite the fact that the network was congested!).

12.6.5 *Conclusions and Future Work*

This prototype is suitable for the infrastructure providers that could not perform over-provisioning and/or re-routing in congested networks. The case is not unrealistic, as it may happen for instance in rural wireless networks. The quality of the service becomes unacceptable and the SLA is not fulfilled anymore by the service provider. As a future work, we envisage the increasing of NC-based GP performance by implementing it on top of MAC Sub-Layer. The processing time being reduced the number of packets lost is expected to decrease to less than 1%. Furthermore the video stream resolution could be increased. With respect to the scalability, whenever a congested link is detected within a network, a six-node butterfly topology could be activated and the cooperative network paradigm described herein could be used. We consider that a network based on butterfly cells, like in a puzzle, is feasible. In such a network NC-based congestion control is useful whenever a more complex solution, like QoS-aware routing, is not available.

12.7 Conclusion

Rather than focusing on a single aspect when implementing theoretical concepts described in previous chapters, the prototyping activities covered multiple, diverse aspects. One major objective of those activities was to combine the complementary concepts also at implementation level. The resulting prototype setups have been successfully demonstrated at different opportunities such as conferences and project presentations. Most of the implementations will be continued and will serve as platforms for future developments (see respective subsections).

References

1. P. Barham, B. Dragovic, K. Fraser, S. Hand, T. Harris, A. Ho, R. Neugebauer, I. Pratt, A. Warfield, Xen and the art of virtualization, in *19th ACM Symposium on Operating Systems Principles* (ACM, New York, 2003), http://doi.acm.org/10.1145/945445.945462
2. P. Bertin, R.L. Aguiar, M. Folke, P. Schefczik, X. Zhang, Paths to mobility support in the Future Internet, in *Proc. IST Mobile Comm. Summit* (2009)
3. T. Biermann, Z.A. Polgar, H. Karl, Cooperation and Coding Framework, in *Proc. IEEE Future-Net* (2009)
4. T. Biermann et al., Description of Generic Path Mechanisms, Deliverable D-5.2.0, 4WARD Project (2009)
5. T. Biermann, C. Dannewitz, H. Karl, FIT: Future Internet Toolbox, in *Proc. 6th International Conference on Testbeds and Research Infrastructures for the Development of Networks & Communities (TridentCom)* (2010)

6. T. Biermann, C. Dannewitz, H. Karl, FIT: Future Internet Toolbox—Extended report, Technical Report TR-RI-10-311, University of Paderborn (2010)
7. T. Clausen, P. Jacquet, Optimized Link State Routing Protocol (OLSR), RFC 3626 (2003)
8. M. D'Ambrosio, P. Fasano, M. Marchisio, V. Vercellone, M. Ullio, Providing data dissemination services in the Future Internet, in *Proc. World Telecommunications Congress (WTC'08)*, New Orleans, LA, USA, 2008. At *IEEE Globecom 2008*
9. C. Dannewitz, T. Biermann, Prototyping a network of information, in *Demonstrations of the IEEE Conference on Local Computer Networks (LCN)*, Zurich, Switzerland, 2009
10. Eclipse Graphical Editing Framework (GEF), http://eclipse.org/gef/
11. N. Egi, A. Greenhalgh, M. Handley, M. Hoerdt, F. Huici, L. Mathy, Towards high performance virtual routers on commodity hardware, in *Proceedings of ACM CoNEXT 2008*, Madrid, Spain, 2008
12. G-Lab Homepage, http://www.german-lab.de/
13. Google: Guice (2007), http://code.google.com/p/google-guice/. Open source project
14. Google: Google Protocol Buffers—Protobuf (2008), http://code.google.com/p/protobuf/. Open source project
15. F. Guillemin et al., Architecture of a Generic Path, Deliverable D-5.1, 4WARD Project (2009)
16. Heterogeneous Experimental Network (HEN), http://hen.cs.ucl.ac.uk
17. INET/INETMANET Framework for OMNeT++ Homepage, http://inet.omnetpp.org/
18. Java New Operating System Design Effort, http://www.jnode.org/
19. R. Koetter, M. Medard, Beyond routing: An algebraic approach to network coding, in *INFO-COM: Proc. of the 21st Annual Joint Conference of IEEE Computer and Communications Societies* (2002), pp. 122–130
20. E. Kohler, R. Morris, B. Chen, J. Jahnotti, M.F. Kasshoek, The click modular router, ACM Trans. Comput. Syst. **18**(3), 263–297 (2000)
21. D. Martin, H. Backhaus, L. Völker, H. Wippel, P. Baumung, B. Behringer, M. Zitterbart, Designing and running concurrent future networks (Demo), in *34th IEEE Conference on Local Computer Networks (LCN 2009)*, Zurich, Switzerland, 2009
22. D. Martin et al., Node Architecture Prototype Homepage, http://nena.intend-net.org/
23. N. McKeown, T. Anderson, H. Balakrishnan, G. Parulkar, L. Peterson, J. Rexford, S. Shenker, J. Turner, Openflow: Enabling innovation in campus networks, SIGCOMM Comput. Commun. Rev. **38**(2), 69–74 (2008), http://doi.acm.org/10.1145/1355734.1355746
24. Mobility Framework for OMNeT++ Homepage, http://mobility-fw.sourceforge.net/
25. G. Nunzi, D. Dudkowski et al., Example of INM Framework Instantiation, Deliverable D-4.2, 4WARD Project (2009)
26. OSGi—The Dynamic Module System for Java, http://www.osgi.org/
27. Packet Capture (PCAP), http://www.tcpdump.org, http://www.winpcap.org
28. P. Papadimitriou, O. Maennel, A. Greenhalgh, A. Feldmann, L. Mathy, Implementing network virtualization for a Future Internet, in *20th ITC Specialist Seminar on Network Virtualization—Concept and Performance Aspects*, Hoi An, Vietnam, 2009
29. C. Perkins, E. Belding-Royer, S. Das, Ad Hoc On-demand Distance Vector (AODV) Routing, RFC 3561 (2003)
30. Z. Polgar, Z. Kiss, A. Rus, G. Boanea, M. Barabas, V. Dobrota, Preliminary implementation of point-to-multi-point multicast transmission based on cross-layer QoS and network coding, in *SoftCOM: Proceedings of the 17th International Conference on Software, Telecommunications and Computer Networks*, Split-Hvar-Korcula, Croatia, 2009, pp. 131–135
31. A. Rus, V. Dobrota, Overview of the cross-layer paradigm evolving towards future internet, Acta Tech. Napocensis **50**(2), 9–14 (2009)
32. G. Schaffrath, C. Werle, P. Papadimitriou, A. Feldmann, R. Bless, A. Greenhalgh, A.Wundsam, M. Kind, O. Maennel, L. Mathy, Network virtualization architecture: Proposal and initial prototype, in *ACM SIGCOMM VISA*, Barcelona, Spain, 2009
33. Rice University, Freepastry (2010), http://www.freepastry.org/FreePastry/
34. A. Varga, Using the OMNeT++ discrete event simulation system in education, IEEE Trans. Educ. **42**(4), 11 (1999)

35. L. Völker, D. Martin, C. Werle, M. Zitterbart, I. El Khayat, Selecting concurrent network architectures at runtime, in *Proceedings of the IEEE International Conference on Communications (ICC 2009)* (IEEE Computer Society, Dresden, 2009)

Chapter 13
Conclusions

Henrik Abramowicz and Klaus Wünstel

Abstract The main conclusions from each chapter are presented, together with a critical perspective. At the end a view on migration paths is given. The deployment of the project results is put into four perspectives: extension of current networks by means of technology, e.g., it is possible to use results from self-management to enhance management capabilities; adding functionality by providing overlay/underlay and control, e.g., the Network of information can be applied on top of the current Internet; network virtualisation is a means of not only sharing network resources, but it can also be used as a migration path to new network architectures e.g. for specialised custom-tailored networks; deployment of a completely new network based on 4WARD's architecture framework, but this has limited commercial applicability and can only be used for very specialised networks, e.g., sensor networks.

13.1 Socio-economics

The success story of the existing Internet shows that progress has not mainly come as a response to requirements arising from the socio-economic and regulatory side, but technical progress has driven adaptation of, e.g., economic models and regulatory rules. As a consequence, there is the need to carefully monitor and assess the outcome from new technical developments, which may initiate new services and applications, and thus will have impact on our society and economy. Due to the mutual influence of technical and non-technical driving forces, all these interactions need

H. Abramowicz (✉)
Ericsson Research, Stockholm, Sweden

K. Wünstel
Alcatel Lucent Bell Labs, Stuttgart, Germany

L.M. Correia et al. (eds.), *Architecture and Design for the Future Internet*,
Signals and Communication Technology,
DOI 10.1007/978-90-481-9346-2_13, © Springer Science+Business Media B.V. 2011

to be done in iterative steps. Drawing the appropriate conclusions from the interdependence of the non-technical driving forces with technical issues is the key for the deployment of Future Internet innovations. The main findings are outlined below.

Migration towards information-centric networks is a key issue. Users are mainly interested in using services and accessing information, not to be aware of the location of the service realisation or the information. The faster, the easier and the safer any type of information will be accessible, the more the Future Internet will have a disruptive social impact on a rapid spread of Internet services through all social, age and educational strata.

There is the need to **design new, advanced connectivity services to be used by humans or 'things'**. The Internet of Things will be an important part of the infrastructure of the Future Internet. The presence of thousands of different network devices will result in a big variety of connectivity requirements and in an enormous wealth of applications.

The opportunity to have a widespread Internet of Things will open new life scenarios where 'things' will take over the responsibility to perform tasks that are today in the hands of human beings. For example, home appliances going out of order will themselves connect to the proper customer centre, cars that need to be repaired connect to the garage to order assistance, or health devices (e.g., pacemakers) contact the medical aid in case of sudden changes in the bearer health conditions.

There will be the **creation and deployment of new types of networks via virtualisation**. The growing complexity, diversity and heterogeneity of networks are major issues in network operation and maintenance. Future networks, employing self-management capabilities, will significantly reduce networking OPEX and CAPEX. Network providers will be able to choose between either to invest in dedicated new physical network resources or to act as virtual network providers and use the physical resources of other providers. Similarly, the need for local customer service and local assistance in physical networks will be partially substituted by the growing number of delocalised software companies that can be based anywhere. The social impact of remote and delocalised business will consist of a reduced migration into cities, thus, preventing all related social problems.

Security and privacy will be improved. In the Future Internet, security, privacy and confidentiality shall be among the key objectives when designing new information sharing concepts. Network security and information are vital for business transactions and for the protection of personal privacy. The confidentiality of communication and related data traffic shall be ensured by prohibiting the listening, tapping, storage or other kinds of interception or surveillance by persons other than the ones involved in the communication without the consent of the users concerned.

The management of information privacy and related responsibilities need to be clear. It will be a major challenge from the legal perspective, on the one hand, to ensure privacy and security of data, and on the other, to keep the Future Internet an open platform for business and administrative applications, entertainment, information exchange, etc.

The great challenge for secure networking applications will be the huge number of objects (several orders of magnitude greater than the number of today's devices or connections) they will have to handle, in both physical and virtual networks.

13.2 Technical Results

The 4WARD System Model and the Architecture Pillars point to the key results of 4WARD and can be understood as 'cornerstones' of what will be a more precise definition of an architecture for the Future Internet. Such an architecture will likely include also other building blocks in order to provide a complete and suited architecture for any type of network that would make up a part of the Future Internet. Below are summarised some of the salient conclusions from technical point of view.

13.2.1 Network Design

In the context of the Future Internet it is envisioned that different network architectures can co-exist and share a common infrastructure. These network architectures can be specifically tailored to particular user or application requirements and, furthermore, can take into account the characteristics of the available networking resources. Therefore, the design of new network architectures is simplified and the productivity of those designing such architectures (called network architects) is expected to increase considerably. Overall this may have a high impact on new innovations on the socio-economic side. Developing network architectures may no longer be a long and painful act of slow-moving standardisation bodies. Moreover, the provisioning of architectural principles, design patterns and building blocks may lead to the deployment of new business cases. VNets provide a versatile platform to ease such a deployment and also lower the barrier of entry from economic point of view.

To model new Network Architectures, some concepts, terms and the basic constructs have been defined. This Architecture Framework provides two levels of views on network architectures:

(i) the macroscopic view mainly focuses on structuring the network at a higher level of abstraction and introduces the concept of Strata as a flexible way to layer the services of the network that can enable the usage of information across different layers; and

(ii) the microscopic view concentrates more on the functions needed in the network nodes, their selection and composition to Netlets that are instantiated in the Node Architecture that allows the dynamic coupling of applications and network protocols not easily possible in today's networks.

The Functional Blocks are presented as the common points between the two views of the architecture. The Component Based Architecture constructs and principles are used as the basis to provide reusable frameworks that minimise the design and development times of new network architectures.

The methodology to be followed is represented in the Design Process; which considers 3 main phases:

(i) detailed requirement analysis,

(ii) the Abstract Service Design and

(iii) the Component Design Phase.

This proposal does not aim to substitute the current design processes available in different organisations but it presents a proposal which links the communication system design and the software development principles. The Design Process is complemented with a Design Repository which provides guidelines, design patterns and the like, in order to support the Network Architect.

Besides looking at the technical design process we have tried to extend this so it should cover the design process from business requirements to a deployable network. This was done by describing a business use case, the Ad-Hoc Community (AdHC), the roles emerging from the business case, possible actors and users in order to extend the design process going from business to deployable networks.

This formed input to define a set of business-level requirements and further on a set of technical requirements. The technical requirements in turn form input to the design process, where through a set of refinement steps which include the design of strata and netlets, as well as a selection of (SW) components, a deployable network architecture was derived.

The step-wise process of going from the macroscopic network-wide components known as strata, and then break them down on the microscopic level to Netlets and SW components, is straight-forward and easily conceivable to be implemented as a tool that could support a significant level of design automation. Still, further research is needed to define a formalised support for the bridging of business level requirements with technical requirements in order to provide a fully seamless step-wise process integrating all the phases of the development cycle from business creation to deployment.

Virtual networks open up a migration path to new network architectures, which can be built according to the previously described Design Process, and also introduce new flexibility for Infrastructure Providers (InPs) with respect to their resource management. The presented VNet framework considers the use of virtual networks in a commercial setting. In contrast to other approaches the architecture considers four involved business roles:

- the Infrastructure Provider,
- the VNet provider,
- the VNet Operator, and
- the VNet end-user.

The framework covers the different necessary signalling and management interfaces that are required during the presented lifecycle of a VNet. Although there are already many virtualisation techniques and mechanisms in wide use today, there is still some work needed in order to realise the fully featured VNet architectural framework. Especially, coordinated interaction between different InPs (e.g., for setting up virtual links) requires standardisation of the respective signalling interfaces. Parts of the presented VNet architecture were realised and evaluated in form of individual feasibility tests.

13.2.2 Naming and Addressing

For the Naming and addressing we have identified five crucial data structures:

(a) a binding table describing which entities have bound themselves to names of other entities (typically in other compartments),
(b) a routing&forwarding table that exists per entity or, as a simplification, per compartment and per node compartment,
(c) the name resolution table, which describes how two neighbouring entities of one comportment actually communicate with each other using which entities in which other compartments,
(d) the idea of a service graph, which contains all possible usage relationships of entities within a node compartment, and
(e) a configuration table for the name resolution process, describing which compartments can attempt name resolution via which other compartment and what parameters are required to do so (with service graph and name resolution configuration closely tied in with each other).

Based on this abstract treatment, we have discussed the process of name resolution in a general fashion. Many communication primitives can be cast into this light, for example, we came to realise that neighbour discovery and name resolution are, at the core, the same thing. We believe that a rigorous treatment of communication design that gives name resolution its proper place across the protocol stacks will lead to a more general, more extensible, and more flexible communication system than what we are currently faced with. It can incorporate automatic discovery of communication opportunities and can solve problems like session mobility quite easily. From a practical perspective, it also can handle a multitude of namespaces and protocol families, without having to standardise on any single namespace across widely differing communication needs.

13.2.3 Security

Security principles are fundamental requirements, but security itself breaks down into implementation choices and tradeoffs involving usability, business models, management, and fundamental understanding of what are needed. Because the future networks considered bring about paradigm change especially regarding the end to end principle and the role of the infrastructure, a bit of caution is needed. Owners and governments will be as eager as before to exert control over resources and usage.

Our current internet security concerns might not be applicable as such when the fundamental architecture has been changed towards a "publish–subscribe" system, as the 4WARD information centric view implies. On the other hand, finding information will be subject to new security challenges where user privacy might not be highly respected.

13.2.4 Interconnection

Interconnection is one of the corner stones in the success of the current Internet.

Despite limitations, it has provided the tools for a sustainable growth in the last 30 years. The paradigm shift implied by the Future Internet has a great impact on the way interconnection has to be tackled. Virtualisation will exacerbate one of the key issues in providing glitch-free Interconnection: traceability for the purpose of commercially sound Interconnection and Service Level Agreements between providers. The provision of multi-domain Quality of Service is one of the new frontiers in Telecommunications in this sense.

In order to make all these challenges feasible, 4WARD has identified a set of principles which will help overcome many of the growth problems currently observed in the Internet. It has also provided the first conceptual building blocks for Inter-domain interconnection, both from a purely architectural point of view and applied to the interconnection of virtualised networking environments. However, the most significant change in the design philosophy of the Future Internet is that Inter-domain Issues are being considered as an integral part of its Design and not as an add-on a posteriori.

13.2.5 Network Management

As networks grow larger and more heterogeneous, In-Network Management (INM) techniques are likely to become significantly more important. This is driven by a need to lower manual intervention to reduce operational expenditure, in combination with a need for scalable solutions to manage ever larger and more complex networks. INM solutions, that are inherently scalable and autonomous, will in the future find applications in all types of networked systems.

The main achievements and results of the work within 4WARD on In-Network Management are the creation of INM framework to support management operations and a set of distributed management algorithms. The INM framework serves as an enabler of management functions. By defining three main elements, namely management capabilities, self-managing entities, and management domains, complex management functions can be constructed and modelled within the framework. The projects prototype implementations illustrate key features that allow to enforce and monitor objectives and to induce self-adaptive behaviour in collaborating management capabilities.

For estimating the network state, which is a necessary input to self-adaptation mechanisms, we have focused on a subset of the management tasks involved in situation awareness. Specifically, we have outlined real-time monitoring of network-wide metrics, group size estimation, data search, and anomaly detection.

Multiple aspects of self-adaptation have also been outlined, including ensuring network stability under control changes and emergent behaviour based congestion control. We have seen that robust, distributed algorithms can be devised for a multitude of management tasks without introducing excessive amounts of overhead in the networked devices.

13.2.6 Connectivity

The Generic Path (GP) model aims to open the path for faster technological convergence in the current Internet and promote functional evolution by introducing object-oriented design concepts in network engineering.

It opts to satisfy the following objectives

- Expressibility of functionality, for describing and designing data communication and network services
- Independently of underlying technology.
- Abstraction of different communication and service paradigms.
- Usability, through a set of primitives that can support generic, selective and polymorphic access to the mechanisms and technology that implement services and service functionality.

From a high-level view a Generic Path refers to a generalised data service path in a network, that data flow across, and in doing so they are subjected to processing and transportation operations.

One important pillar of the Generic Path approach for the clean-slate design of the future Internet is a suitable mobility concept for complex communication scenarios and new upcoming services. We have introduced several mobility approaches ranging from dynamic mobility anchoring, anchorless mobility, naming and addressing schemes to multi-path transfer and enhanced content delivery networks schemes. All these approaches seem promising and fulfil major requirements concerning flexibility and scalability. By combination of the proposed concepts we get the following advantages: First the combined mobility approach works for all kinds of mobility, be it session, terminal or network mobility. Second, it works across a variety of situations, be it slow or fast moving users in small hotspots or big macro cell environments.

Unlike today's tunnelling solutions, the combined mobility concept allows local routing, flexible anchoring and multi-homing. Thereby it leads to decreased transmission delays and lower bandwidth utilisations and improved user experience.

13.2.7 Information Objects

We have taken the information centric paradigm as the basis for our work.

From this we have developed an information model encompassing not only virtual data objects but also real world objects as well as services. To make it possible to design a new information centric network architecture that is more scalable and has better security properties than today's Internet architecture one key component needed is a new naming scheme. Through the design of our naming and security framework we have been able to build an architecture that provides security for the information objects themselves rather than for the boxes containing them and the links interconnecting the boxes.

The information-centric communication abstraction has a number of advantages. Efficient distribution of content to a large set of recipients can be implemented.

Content caching is built into the architecture and is thus provided without resorting to either interception of requests or special configuration at the receiver. Load-balancing is provided without depending on add-ons such as DNS round robin.

Both reliability and performance are improved, since information can be retrieved from the closest available source.

Performance and reliability can be enhanced by an information-centric paradigm in a heterogeneous wireless environment where there are disruptions in communication, transient access opportunities and multiple access choices. An information-centric communication abstraction gives more flexibility in the delivery of data objects compared to using an end-to-end byte-stream. The network has better knowledge of the intent of the applications and therefore has the possibility to treat the data more intelligently. The network can easily deliver the data using multiple routes, redundancy over the available paths, and intermediate storage to overcome connectivity disruptions. The extreme is a scenario when an end-to-end path never exists. With an end-to-end reliable byte-stream, the network has to violate the assumptions of the abstraction, or the application must implement these functions itself, including application gateways at suitable network locations.

The performance and reliability benefits from using distributed storage in the NetInf nodes also benefit non-dissemination applications, such as personal email. Another example is that direct delivery of email between two laptops with WiFi connectivity can be supported without involving infrastructure.

The information-centric approach gives new possibilities to prevent denial-of-service (DoS) attacks. With an information-centric approach nobody can force network traffic your way without your consent. This is made possible by moving control from the sender to the receiver. A sender can make information available, but for that information to be transferred, the receiver has to ask for it. Prevention of DoS attacks is a motivation to design NefInf as a replacement to the current TCP/IP for end-to-end communication.

13.2.8 Prototypes

The 4WARD project developed a number of prototypes to demonstrate feasibility and proof of concepts in a number of different areas. Besides these goals we also wanted to create a SW base that is open for others to experiment and further develop our concepts. Below are summarised some results and proposed further work regarding prototypes about network design, virtual networks, self-management, Generic Path and Netinfo. It should be noted that some of the code used for the prototypes will be released as open source software.

The prototype implementation of the **Node Architecture Daemon** including a simple example architecture resulted in a demonstration shown at the Local Computer Networks conference 2009 and was awarded the best demo based on attendee voting. Using a simple video application, we showed how we mitigated the impact of degrading network conditions leading to a deteriorated service quality. Measures could be taken either in the application or in the communication protocol in order to cope with the changed situation. Rather than adapting the application or introducing complex, adaptive protocol features, we showed how we re-designed and enhanced the simple Netlet, used before the deterioration, with further protocol mechanisms using the Netlet Editor.

The enhanced Netlet was then deployed on the Node Architecture and the selection process decided to use the new Netlet since it provided a more robust data delivery. The demonstration setup has been enhanced with a small virtualisation test-bed where the video streaming scenario was seamlessly transferred into a virtual network. The virtualisation test-bed, shown at several project-wide events, illustrated the roles of the different stakeholders and was able to show the feasibility of resource reservation and isolation of virtual networks.

The key features implemented by this prototype setup include:

• Rapid creation of composed protocols (Netlets) using a graphical design tool
• Reuse of existing protocol building blocks
• Automatically selecting an appropriate Netlet based on application requirements and network properties experienced at end nodes
• Isolation of virtual networks
• Separation of roles in virtual networking scenarios.

To broaden the basis and get more diverse results, our future plan is to include attachment of end systems to multiple virtual networks, and to test the daemon also in mobile ad-hoc scenarios. Some aspects will be further developed within G-Lab http://www.german-lab.de/, a German test-bed platform distributed among multiple locations. The code of the Node Architecture prototype is released under the conditions of the GNU General Public License (GPLv2) and is available online https://i72projekte.tm.uka.de/trac/nodearch/.

Another prototype **Vnet Management Environment** that was implemented enables the provisioning and management of customised VNets on top of multiple InPs, in accordance with the network virtualisation architecture described in this book (see Sect. 4.9). This prototype also provides insights into critical architectural design decisions, such as the tussle of information disclosure against information hiding.

Despite the increased complexity for resource discovery and allocation, we used this prototype to show that VNets can be provisioned within a few seconds. Our experimental study did not uncover any scalability issues within our infrastructure, even in the presence of multiple InPs.

With our prototype implementation, we also showed what technological ingredients are needed and how these can be combined efficiently to provision and manage VNets. The main building blocks of our prototype, such as node virtualisation technologies and packet encapsulation/decapsulation for virtual link setup, are readily available today. Our results indicate that with this technology the barriers of entry into the operator market has been lowered and with new business models commercial products can easily be created. This shows that a shift towards full network virtualisation is viable in the not too distant future.

The **real-time adaptation in the emergency** prototype has demonstrated self-organising and self-deployment of a network domain through the holistic and systematic approach of the strata architecture. The need for advanced management instruments that can tackle the complexity in operating the network, both for the technical aspects related to network performances as well as non-technical aspects related to operator's objectives has been demonstrated. The INM framework and algorithms for monitoring and congestion control provide distributed domain level congestion detections with optional self-healing capabilities while also allowing domain owners manage their domains at business objective level. Finally interoperability via SLA negotiation at a domain level is highlighted in the prototype with the SLA switched networking enabled via Strata encapsulation of the different network architectures at a domain level.

Both the **NetInf and the GP prototype**s prove the benefits of the architectural concepts that have been introduced in Chaps. 9 and 10. In particular, the NetInf prototype demonstrates the specific NetInf advantages like improved resource usage, inherent load balancing, efficient data dissemination, and improved security features that are deeply integrated into the network architecture. In addition, the GP prototype shows that disposing of the strict ISO/OSI layering provides a lot of flexibility. It permits to easily implement networking features that are difficult to achieve today, e.g. session mobility and dynamic use of cooperation and coding techniques, which require a lot of cross-layer information. Integrating both prototypes has significantly improved the understanding of how both concepts interact and how they can benefit from each other. Two examples described in Chap. 12, bootstrapping and session mobility, illustrate the potential of GP's generic name resolution scheme and of building custom-tailored GPs that provide NetInf-specific transport. Future examples may include GPs that support transparent in-network caching.

We have released the NetInf and GP prototype code base as open source project to make it available to a wider audience.

The prototype on using **INM in GP for cross-layer QoS** is suitable for the infrastructure providers that cannot perform over-provisioning and/or re-routing in congested networks, which can happen in rural wireless networks. The quality of the service might become unacceptable and the SLA is not fulfilled anymore by the service provider. As a future work, we envisage the possibility to increase Network Coding(NC)-based GP performance by implementing it on top of MAC Sub-Layer. The reduction of the number of packets lost is expected to decrease to less than 1%. Furthermore the video stream resolution could be increased. With respect to the scalability, whenever a congested link is detected within a network, a six-node butterfly topology could be activated and the cooperative network paradigm described herein could be used.

Rather than focusing on a single aspect when implementing theoretical concepts described in previous chapters, the prototyping activities covered multiple, diverse aspects. One major objective of those activities was to combine the complementary concepts also at implementation level. The resulting prototype setups have been successfully demonstrated at different opportunities such as conferences and project presentations. Most of the implementations will be continued and will serve as platforms for future developments.

13.3 From Research to Reality—Migration Paths for Future Internet

4WARD has used an approach of clean slate research meaning that we have conducted the research without any constraints related to the current networks. We wanted to attain more optimal approaches without being impeded by compatibility issues at the very start. This on the other hand poses the issue of how to use the research results and apply/migrate it to current networks in order to evolve them. A rollout of a complete new network with novel architecture is generally not really economically viable.

There are different possibilities to deploy the results of 4WARD and we have basically the following options

- Extend the current networks by means of technology e.g. it is possible to use results in self-management to enhance the management capabilities by adding monitoring algorithm that is not specific to any service. GAP is such an example that is providing monitoring.
- Add functionality by providing overlay/underlay and control. An example of overlay is e.g. Netinf that can be applied on top of the current Internet or adding functionality that is orthogonal to the current network. An example the latter is the design process defined in Chap. 4 that also could be used to customise networks. Another example is making use of network coding in existing wireless structure as an example of underlay.
- Network virtualisation is a means of not only sharing network resources but can also be used as a migration path to new network architectures e.g. for specialised custom-tailored networks. An obvious approach is to apply this to more confined segments like certain enterprise segments.
- Deploy a completely new network based on 4WARD architecture framework that is separate and provide gate-waying to the current networks but this has limited commercial applicability and can only be used for very specialised networks e.g. sensor networks. Another example would be the deployment of research networks at large, which could be implemented along existing infrastructure and could make use of real world traffic. In the US there are similar plans for GENI. Protocols and mechanisms could be tested without disturbing the "real" network.

Looking at the results of 4WARD we can see that from a network design point of view the whole design process and tools like design repository can be applied since it is orthogonal.

The **network virtualisation** is probably **the tool** to use for migrations as the we have currently a trend to make use of virtualisation in general like cloud computing in order for users to benefit from a more linear cost curve that is a result of large scale usage and elasticity.

The economic aspect of network virtualisation lies in the optimisation of needed resources and therefore in reduced Total Cost of Ownership (TCO). For different kinds of virtualisation, there are also additional advantages, such as license sharing or power reductions in server virtualisation. The point is that each virtual network can be built according to different design criteria and operated as a service-tailored network while running together with a big variety of other virtual networks on the same infrastructure.

The costs for enabling migration have to be considered carefully. Typically, operators deploy more and more networks often on the same infrastructure. In an ideal case, customers switch from one platform to another and operators could turn off "old" platforms. In reality, this is much more complex. Service interruption is a major pain point for customers. It could be avoided only in a small time frame and the number of changes in this time frame is rather limited. In a large scale network, it requires a lot of time and planning in order to migrate. On the other hand, operators are not in favour of terminating contracts with customers and setting up new ones. So migrating to a platform that has virtualisation already in place would be very beneficial: as the service virtually is maintained, there is no need to change any contract. But there remains from an operator perspective an important drawback, that it seems that there is no simple way to migrate to virtualisation otherwise than installing new platforms (with virtualisation capabilities integrated) and shifting the majority of customers together with their currently used services to the new platform while only a small portion of customers will take the chance to switch to an up-to-date service by active migration.

From self-management point of view or INM the framework allows for positioning Management Capabilities, at different levels, namely:

- inherent—very tightly coupled with the entity being managed, e.g. TCP flow control mechanisms in today's networks. The management is part of the protocol.
- integrated—coupled with the entity being managed, e.g. ANR (Automatic Neighbour Relation) functionality in LTE. This detects and configures the relationships between cells. It is part of the cell management.
- separated—decoupled with the entity being managed, e.g. a monitoring algorithm which is not specific to any service but which a service can make use of. The GAP algorithm is an example of this in that it provides a monitoring of whatever parameters are needed.
- external—completely external, more so non-INM management, e.g. the OSS functionality today is external to the network itself. Inter-domain management may still need to reside at this level.

Both the integrated and the separated approaches can be seen as possibilities to apply results to the current networks and by this evolve them into more self-management.

Generic Path as defined focusing on APIs is easily compatible with the current network protocols. One can even see some possibilities of enhancement to the Open flow approach that could benefit from the GP concept.

Also network coding can be applied e.g. on top of a MAC to increase bandwidth on congested links.

The Netinf approach can be launched on top IPV4 or e.g. IPV6 although it can benefit more from running on top of GP.

Many of the approaches described in the book will be progressed by new projects, be proposed to standardisation and/or be put in open source for other to take on where the project has left it.

Appendix
Project Description and Reports

The project 4WARD started January 2008 and was completed by June 2010. It was partly funded under the EU Framework Programme 7, with a total budget around 24 million euros, from which the European Commission funded almost 15 million euros.

The project consisted of the following partners:

- Alcatel–Lucent, Germany
- Deutsche Telekom, Germany
- Ericsson, Canada
- Ericsson, Finland
- Ericsson, Germany
- Ericsson, Sweden
- France Telecom, France
- Fraunhofer Gesellschaft zur Förderung der angewandten Forschung, Germany
- Fundación Robotiker-Tecnalia, Spain
- Groupe des Ecoles des Télécommunications, France
- Instituto de Telecomunicações de Aveiro, Portugal
- Instituto Superior Técnico—Technical University of Lisbon, Portugal
- Karlsruhe Institute of Technology, Germany
- Kungliga Tekniska Högskolan, Sweden
- Lancaster University, United Kingdom
- NEC Europe, Germany
- Nokia Siemens Networks, Finland
- Nokia Siemens Networks, Germany
- Portugal Telecom Inovação, Portugal
- Rutgers University, United States of America
- Siemens, Romania
- Swedish Institute of Computer Science, Sweden
- Technical University Berlin, Germany
- Technion—Israel Institute of Technology, Israel
- Telecom Italia, Italy
- Telefónica Investigación y Desarrollo, Spain

L.M. Correia et al. (eds.), *Architecture and Design for the Future Internet*,
Signals and Communication Technology,
DOI 10.1007/978-90-481-9346-2, © Springer Science+Business Media B.V. 2011

- Telekomunikacja Polska, Poland
- Universitatea Tehnica din Cluj-Napoca, Romania
- University of Basel, Switzerland
- University of Bremen, Germany
- University of Paderborn, Germany
- University of Surrey, United Kingdom
- Valtion Teknillinen Tutkimuskeskus, Finland
- Waterford Institute of Technology, Ireland

The project was structured into 6 Work Packages as follows:

- WP1—Business Innovation, Regulation and Dissemination (BIRD)
- WP2—New Architectural Principles and Concepts (NewAPC)
- WP3—Network Virtualisation (VNet)
- WP4—In-Network Management (INM)
- WP5—Forwarding and Multiplexing for Generic Paths (ForMux)
- WP6—Network of Information (NetInf)

The majority of the project reports (Deliverables) were public, which are listed below:

- D-0.1 Dissemination and Exploitation Plan
- D-0.4 ATF Report
- D-0.5 Final Report
- D-1.1 First Project-Wide Assessment on the Non-technical Drivers
- D-1.2 Evaluation of 4WARD Business Use Cases
- D-1.4 Final Assessment on the Nontechnical Drivers
- D-2.1 Technical Requirements
- D-2.2 Draft Architectural Framework
- D-2.3.0 Architectural Framework: New Release and First Evaluation Results
- D-2.3.1 Final Architectural Framework
- D-3.1.1 Virtualisation Approach: Concept (Final)
- D-3.2.0 Virtualisation Approach: Evaluation and Integration
- D-3.2.1 Virtualisation Approach: Evaluation and Integration—Update
- D-4.1 Definition of Scenarios and Use Cases
- D-4.2 In-Network Management Concept
- D-4.3 In-Network Management Design
- D-4.4 In-Network Management System Demonstrator
- D-4.5 Evaluation of the In-Network Management Approach
- D-5.1 Architecture of a Generic Path
- D-5.2 Mechanisms for Generic Paths
- D-5.2.0 Description of Generic Path Mechanism
- D-5.3 Evaluation of Generic Path Architecture and Mechanisms
- D-6.1 First NetInf Architecture Description
- D-6.2 Second NetInf Architecture Description
- D-6.3 NetInf Evaluation

All the relevant information concerning the project can be found at http://www.4ward-project.eu.

Glossary

Chapter 2

4WARD Tenet Major Objectives to be achieved by 4WARD

Architecture Framework A framework defining a possible architecture for the Future Internet

Design Process Definition of all the necessary steps to define a new architecture

System Model Model defining all the necessary blocks, as defined by 4WARD, for the creation of Future Internet

Chapter 3

Business Model Definition of the business environment for a technology and/or application, describing step-by-step the area where the business is inserted

Costumer Player using the services provided by the operator

Environment Variables that enable the creation of a business model

Interconnection Connection agreements between different providers

Migration Steps that must be taken in order to upgrade something, e.g., steps that must be taken in order to upgrade current networks to the Future Internet ideas

Non-technical requirements Guidelines that must be observed in order to allow the development of the Future Internet

Physical Network Provider extracts a part or the entire virtual network from its own physical resources and binds them on behalf of the virtual network providers for later use by the end users and service providers

Privacy Ability for someone to hide his own personal information, or to not allow for some to make it public without his knowledge

Scenario Hypothetical situation defined in order to evaluate the development of new applications, technologies and related business environment

Service Provider deploys services or applications on virtual networks

Services Applications and technologies provided to customers

Use-case Complete business model for and hypothetical situation, for given services or applications

Value chain Chain of activities that must be fulfilled in a given business

Virtual Network Operator operates, maintains, controls and manages the virtual network

L.M. Correia et al. (eds.), *Architecture and Design for the Future Internet*,
Signals and Communication Technology,
DOI 10.1007/978-90-481-9346-2, © Springer Science+Business Media B.V. 2011

Virtual Network Provider requests virtual resources from various infrastructure providers

Virtual Network User end customer accessing applications over virtual networks

Chapter 4

Component Based Architecture The decomposition of the engineered systems into functional or logical blocks with well-defined interfaces used for communication across these components. Components are considered to be a higher level abstraction than objects and as such they do not share state and communicate by exchanging messages carrying data

Contract A contract is the unit of interoperability. It represents the meta-data and specification that is attached to an interface that mutually bind clients and providers (implementers) of that interface

Design Pattern A design pattern provides a scheme for refining elements of system or relationships between them. It describes a commonly recurring structure of interacting roles that solves a general design problem within a particular context

Design Repository The Design Repository is a focus point for Stratum, Netlet and Component Based Architecture within the overall process. This repository is where the linkages between the concepts and overall knowledge are stored. The repository does not stand in isolation, it is constantly updated as a result of feedback during the execution of the process—it learns and adapts to new patterns, functions, and architectures as they are designed and created by the designer

Design The activities performed by developers that result in the architecture of a system. The term is also used as a name for the result of these activities

Functional Block A sequence of instructions that realises a certain functionality protocols and other functions can be built with. Commonly only local functionality, which makes up the distributed functionality by communicating with Functional Blocks in components on other nodes. Also called Building Block. A simple example of a Functional Block would be the calculation of a Cyclic Redundancy Check, which belongs to the functionality error control

Horizontal Stratum Horizontal Strata provide the resources and capabilities for communication and information management across networks

Interface An interface is a list of signatures. A signature describes an abstract function, which may either be offered or required to access functionality. Signatures typically encompass the following: name, return types, ordered list of parameters with types, (optionally) set of possibly thrown exceptions, (optionally) pre-conditions and post-conditions (design by contract)

Netlet Selector The Netlet Selector contains an automated selection approach which chooses the best Netlet for a given task

Netlet Netlets are components of the node architecture that contain a local collection of functional blocks to realise a set of protocols in a specific network architecture

Network Access A Network Access provides an interface to access any underlying network infrastructure for a Netlet

Node Architecture The Node Architecture proposes an internal structure of a node and how to select, instantiate, and run different Netlets of arbitrary network architectures in parallel

Protocol A protocol describes the syntax, semantics and functional behaviour of the message-passing mechanism between components

Stratum Gateway Point A Stratum Gateway Point provides access to other strata being of the same or similar type (i.e. having a common point of origin regarding its specification), but independently realising functionalities in different networks. Stratum Gateway Points are points where strata may interoperate if necessary. A Stratum Gateway Point is further decomposed into one or more Interfaces

Stratum Service Point A Stratum Service Point provides access to the capabilities and functions of a Stratum. A Stratum Service Point is further decomposed into one or more Interfaces

Stratum The stratum is a structural element of a network architecture for designing, realising and deployment of distributed functions in a communication system

Substrate link A substrate link is a physical link—wired or wireless—interconnecting two substrate nodes. For the provisioning one has to discriminate between substrate links under exclusive control of one infrastructure provider and substrate links between infrastructure providers. In the first case, both substrate nodes terminating the link are owned by the same infrastructure provider; in the second case, the terminating substrate nodes belong to different infrastructure providers

Substrate node The term substrate node refers to physical hardware that is owned by infrastructure providers and exclusively controlled and administered by them. Infrastructure providers may decide to offer a virtualisation service on virtualisable substrate nodes to other parties, e.g., VNet providers and operators and assign shares of their hardware to them. Non-virtualisable substrate nodes cannot be used for this purpose, e.g., as functions required for virtualisation are not implemented in them. Nevertheless, these nodes might be part of a virtual link and might therefore have to be support virtual networks, e.g., to enable QoS guarantees for a virtual link

Substrate The substrate consists of all substrate nodes and substrate links. The Design Repository is the repository where the architectural principles and design patterns are stored together with the specific design decisions and the result of the design. The functionality of a Stratum is encapsulated by and offered as a service to other Strata via a Stratum Service Point. A stratum implementing the same or similar distributed function but in a different domain (thus assuming independent implementations) can interoperate via a Stratum Gatewaying Point

Vertical Stratum Vertical Strata are those responsible for listening and ordering other strata; listen to KNOW about the network in order to GOVERN it

Virtual link A virtual link can be a reserved share of a physical link or can be composed from multiple physical links, e.g., by splitting a virtual link between two virtual nodes over multiple substrate links or by spanning a virtual link over multiple substrate links. Depending on the link technology, e.g., IEEE 802.11 or MPLS, there are already existing mechanisms to support virtualisation that can then be employed to enforce separation and isolation between virtual links

Virtual network slice A virtual network slice consists of the reserved resources for virtual nodes and virtual links belonging to a virtual network, i.e., the virtual node slices and unused—but reserved—virtual links

Virtual network topology A further discrimination can be made when taking end users into account. To explicitly refer to a virtual network without considering end users' devices that are not under administration of an infrastructure provider, the term virtual network topology can be used

Virtual network A virtual network is a running instance of a virtual network slice. This implies configured and active virtual nodes as well as virtual links that are potentially in use

Virtual node slice A virtual node slice is a set of resources that has been reserved on a substrate node. In contrast to a virtual node, a virtual node slice is not yet executing any code and is therefore only an intermediate form of existence towards a living virtual network

Virtual node A virtual node is built from a virtual node slice by installing and booting an operating system and by setting up any required applications

Chapter 5

Address binding An agreement between two entities that one is reachable via the other and that one entity's name serves as the other's address. For a binding to be possible, both entities must share (at least) one compartment. Typically, this is a node compartment

Address An address of entity 1 is a name of entity 2, where entities 1 and 2 have agreed on a binding between each other. Hence, entity 1 becomes reachable via this address and via entity 2's compartment

Binding table A node compartment can choose to implement name/address bindings such that they are all collected in a node-wide table, called the binding table. Alternatively, implementations which only store these bindings in the rejective entities are possible

Binding A relationship between two entities sharing (at least) one compartment, typically a node compartment. A binding implies that not only is it possible for these two entities to communicate, one of them has agreed to accept traffic on the other's behalf and pass it on to the other

Compartment A compartment is a functionally complete communication system, restricted by scope. It provides a communication service and typically uses other compartments' communication services. It is defined by (a) a set of entities, (b) a namespace, (c) mutual reachability among all entities inside the compartment. Entities inside a compartment must understand the compartment's namespace and set of protocols

Forwarding table A compressed version of a routing table, containing, for the known destinations, only the next-hop neighbour in the given compartment as well as how this neighbour shall be reached (i.e., a pointer into the name resolution table)

Forwarding The process of consulting the routing table (or a compressed form of it, the forwarding table) to determine the next hop neighbour inside the given compartment to pass a packet (or a circuit) on to

Generic path A generic path is a generalisation of data transport over and/or data transformation inside a physical or virtual network. The term may refer to the architecture as a whole, a specific class of such data transport/transformation paths, or a concrete instance of such a class. Generic paths exist between entities inside a compartment; a generic path instance never leaves a compartment. The set of supported generic path classes is one (of several) defining properties of a compartment

Name resolution configuration table A table containing information how the names of one compartment can be resolved via a given assisting compartment. In complex name resolution configurations, a third compartment may be used to provide storage for name/address bindings. The content of this table can be pre-configured or can be filled in at runtime, using routing inside the compartment looking for name resolution and recursive name resolution

Name resolution table A table containing, for each entity, all the neighbours (as far as they have been already found) of this entity inside the entity's compartment, as well the information how to reach this neighbours: via which assisting compartment, and via which assisting local and remote entity (identified by their names). Cost information can be provided as well

Name resolution The process of finding out, for a given name, a set of addresses (along with the compartments of these addresses) via which an entity with this name can be reached. A compartment can define name resolution to pertain to unicast, anycast, multicast, or other operations

Name An element of a namespace. When an entity obtains a name, the name can be used to identify the entity, but uniqueness is not in general required. Entities may have zero, one, or more names from a given namespace; they may also belong to multiple namespaces

Namespace A set of names along with operations on names. A namespace must at least define the operation "equality" for any two of its elements (it must carry an equivalence relation). A namespace is free to define further operations (e.g., "subname" and "aggregate")

Neighbour discovery The process of finding out, for a given entity E and given compartment C, what are the C neighbours of E and the C neighbours of E with respect to some other compartment C'. The process of neighbour discovery highly depends on C and on C', is closely related to resolving a wildcard name in C, and is hence configured via the name resolution configuration table

Neighbour Being a "neighbour" is a relation between two entities $E1$ and $E2$ inside one compartment $C1$ and is only defined when either $E1$ and $E2$ can communicate directly with other using primitive means of $C1$ or when they can communicate with the help of some assisting compartment $C2$. In the first case, we call $E1$ and $E2$ $C1$ neighbours (but usually drop this for convenience). In the second case, for $E1$ and $E2$ to be neighbours, there must be some other compartment $C2$ with entities $E1'$ and $E2'$ such that $E1'$ and $E2'$, $E1$ and $E1'$, and $E2$ and $E2'$ can communicate with each other (either via $C2$ or via some node compartments). Then, $E1$ and $E2$ are called $C1$ *neighbours with respect* to $C2$. The neighbourhood of $E1$ depends on both $C1$ and $C2$

Node compartment A specific compartment, defined by the boundaries of a computing system (be it a physical or a virtual one). The namespace is the set of process/thread identifiers, the communication protocols are the interprocess communication facilities of the operating system. Multiple virtual machines running on a single physical computing system form several node compartments, as the OS interprocess communication is not available between entities of these different virtual machines

Routing table A routing table contains destination names and neighbour names inside a given entity's compartment, along with estimated costs to reach a destination via a given neighbour. It is allowed for a neighbour name to appear multiple times if it is reachable via different assisting compartments; then, estimated costs to this neighbour (and ensuing routes) might differ

Routing The (usually distributed) process of computing a routing table at some or all entities of a compartment. It is based on neighbourhood information

Service graph The service graph exists per node compartment, has entities as nodes and an edge between two entities *E1* and *E2* if *E1* offers a service useful to *E2*. It is constructed by neighbour discovery in the node compartment and routing on this service graph is a means to recursively construct communication relationships by identifying the required services

Chapter 6

Bluetooth Open wireless protocol for exchanging data over short distances. Implementation related security problems are known

Hypervisor Also called virtual machine monitor. A piece of software or hardware allowing multiple operating systems to run simultaneously in isolation, protecting sensitive areas (memory, application interfaces)

RFID Radio-frequency identification. A collection of technologies, consisting of interrogators (also known as readers), and tags (also known as labels): active RFID tags, which contain a battery and can transmit signals autonomously, passive RFID tags, which have no battery and require an external source to provoke signal transmission, and battery assisted passive, which require an external source to wake up. Commonly used, without regard to serious security and privacy problems

SIM/USIM Subscriber Identity Module, a logical entity often embedded on a Smart Card, used to identify a subscriber on mobile telephony devices and computers. SIM cards are mandatory in GSM devices, in UMTS it is called USIM or the Universal Integrated Circuit Card, which runs a USIM application. Secondary secret keys generated when the USIM interacts with an operator can be used as security anchors for other internet usage, e.g. computers and OpenID

Skype A software application that allows users to make voice calls over the Internet, using a proprietary protocol (encryption cannot be disabled, and is invisible to the user)—the largest international voice carrier. Skype provides an uncontrolled registration system for users, with no proof of identity

WLAN Wireless Local Area Network, often referring versions of the IEEE 802.11 standard (Wi-Fi). Used for local networking replacing cabled LAN, but is also usable for ad hoc networking without a base station; security on the radio level is either lacking, or based on broken WEP, or on newer WPA and WPA2

Zigbee A new suite of high level communication protocols based on the IEEE 802.15.4 standard, to replace WLAN and Bluetooth

Chapter 7

Autonomous System IP network that has a single management authority and is managed using a consistent set of policies

Domain Network Partition based on a common technology, which can operate independently from the rest of the network

Interworking Ability to exchange data

Peering The interconnection of two domains

Quality of Experience Overall acceptability of an application or service, as perceived subjectively by the end-user, including the complete end-to-end system effects (client, terminal, network, services infrastructure, etc.). (from ITU-T G.1080)

Quality of Service The collective effect of service performance which determine the degree of satisfaction of a user of the service. (from ITU-T E.800)

Service Level Agreement Formal agreement between two or more entities that is reached after a negotiating activity with the scope to assess service characteristics, responsibilities and priorities of every part. A SLA may include statements about performance, tariffing and billing, service delivery and compensations. Every performance reporting may include only the QoS parameters agreed in the correspondent SLA (from ITU-T E.860)

Chapter 8

Anomaly detection Analysis of measurements deviating from normally observed behaviour

Capital expenditure (CAPEX) Cost of acquiring or upgrading physical assets such as networking equipment

Co-design Style of designing management functions in conjunction with service functions

Collaborative fault-localisation Isolation of abnormal behaviour to certain network components

FCAPS Fault, configuration, accounting, performance, and security. A model and framework for network management

GAP Generic Aggregation Protocol. Distributed algorithms that provide continuous monitoring of global metrics

Global management point High-level entry point via which a network is managed in terms of high-level objectives and according to the INM paradigm

INM Framework Set of architectural elements and concepts supporting INM algorithms and management functions

INM In-network management. Performing management tasks within the network itself

Management capability The building blocks for composing any basic and any more complex management functions from management algorithms

Management domain Specific view on a set of self-managing entities, either structural or functional, providing access to a restricted set of management functions only

NATO! Not all at once, a statistical scheme and algorithms for precisely estimating the size of a group of nodes affected by the same event, without explicit notification from each node, thereby avoiding feedback implosion

Network situation awareness Monitoring and understanding of network performance

Operational expenditure Ongoing cost for running e.g. a network

Self-adaptation control loop An algorithm or a portion of an algorithm within a Management capability that implement self-adaptation functionality for an INM algorithm

Self-adaptation Network management actions taken by the network itself to adjust to changing conditions

Service access point Service access point. High-level entry point into management services within the 4WARD network management framework

Chapter 9

Compartment Set of Entities that fulfil some requirements (each Entity carries a name from a Compartment specific name space, all Entities in a Compartment can communicate, all entities in a CT may communicate)

EndPoint It is a thread or process executing a data transfer protocol machine and doing any kind of traffic transformation

Entity Generalisation of an application managing a resource

Flow Sequence of datagrams that share certain properties

Generic Path: A Generic Path is an abstraction of data transfer between communicating Entities located in the same or in remote nodes

Hook A Hook is a Generic Path within the node compartment

Path Route used by the information to get from the point of origin, to the point of destination

Routing The act of path establishment in a distributed system

Chapter 10

Identifier–locator represents the logic association between the Identifier of an Information Object and the address (locator) where the effective payload (usually a Bit level Object copy) resides

Information Binding represents the action of connection the Information Object to a Bit-level Object

Information centric paradigm In communication history this paradigm represents the current step of evolution after circuit switching and packet switching. This paradigm overcomes the semantic overload of URI's addressing, and aims to create an internet space based on information content, not information addresses

Information Object represents a collection of metadata describing an individual piece of content (stored as Bit-level Object), the locator of a BO, and any other descriptions provided by publisher (security information, for example)

Metadata represents a unitary piece of description associated with an Information Object

Name Resolution System represents the abstraction associated with the Resolution System based on object names instead of object address

Naming Scheme represents the procedure to generate certain object names based on schemes or predefined policies

Network of Information represents in 4WARD the realisation of the Information Centric Paradigm

Self-certification represents the capability of an Information Object to certify about his origin and validity based on metadata content, without the usage of an independent certification authority

Chapter 11

Business model It describes how an organisation creates, delivers, and captures economic, social, or other forms of value. In the present paper it is used to extract value from 4WARD innovations. It connects the world of technical experts and their technical inputs with the world of business experts to produce economic outputs

Business use case It describes a business process from an external, value-added point of view, possibly including partners and suppliers, in order to provide value to a stakeholder of the business. Useful to understand or change business processes

Incentives It is any factor that enables or motivates a particular course of action, or counts as a reason for preferring one choice to the alternatives. It is an expectation that encourages stakeholders to behave in a certain way

Information object Any type of data that can be delivered on a network

Interface The point of interaction or communication between two or more hardware or software entities

Player An entity that is able to play a specific role in a business model

Role Each opportunity of exchanging tangible or intangible goods or money that can be identified to develop a business model: for each role several players could be viable

Stakeholder Any person, group or entity that has direct or indirect stake in an organisation because it can affect or be affected by the organisation's actions, objectives, and policies. Typically they include customers, employees, owners (shareholders), suppliers of any type

Win-win model It is a business model, based on the mathematical game theory, which is designed in a way that all participants can profit from it in one way or the other

Chapter 12

Component-based Architecture OSGi based architecture for componentised protocol stacks

Generic Path The 4WARD approach of a new network transport architecture

Governance Stratum, Knowledge Stratum Management and monitoring planes of the stratum concept

Information-Centric Network An information-centric network primarily addresses and handles objects or pieces of information, independent of their actual location

Infrastructure Provider Operator and provider of a physical infrastructure

In-Network Management The 4WARD approach for self-management in future networks

Netlets Containers of protocols or protocol stacks that can be loaded dynamically into the Node Architecture

Network architect A person or group of persons who designs network architectures and makes important design decisions

Network Coding Dynamic (re-)encoding of data flows within a network to allow for better adaptations of the actually transferred data to fit to current network constraints

Network of Information The 4WARD approach of a content-centric networking paradigm

Network virtualisation Sharing of physical network infrastructure by different networks while those networks are isolated against each other

Node Architecture An extensible architecture for network nodes (end-systems and intermediate systems) that allows to access multiple networks of different network architectures in parallel, while providing a single, requirement-based application interface; protocols and protocol stacks are loaded as so-called Netlets

Stratum Concept providing general abstractions of network functionality and interfaces

Virtual network A network running on virtualised resources

VNet Management describes necessary tasks and tools for configuring and deploying virtual networks

VNet Operator The network operator of a virtual network who has contracts with VNet Provider

VNet Provider Stakeholder who offers virtual network resources obtained from multiple Infrastructure Providers to VNet Operators

Index

L.M. Correia et al. (eds.), *Architecture and Design for the Future Internet*,
Signals and Communication Technology,
DOI 10.1007/978-90-481-9346-2, © Springer Science+Business Media B.V. 2011